Genre Networks

This innovative book employs genre as a fruitful lens for exploring the complexity of science communication online and the new genre assemblages formed at the interface of multiple genres in digital environments.

Pérez-Llantada and Luzón argue for a conceptualization of Science 2.0 that views digital genres in conjunction with other genres, accounting for the ways in which diverse Internet users choose different points of entry for accessing information on science of varied depth, views, and perspectives. Taking Swales's conceptualization of forms of genre collectivity as its point of departure, the book puts forward this new understanding of multisemiotic genre assemblages in digital science communication, considering dimensions of hypertextuality, intertextuality, and multimodality in the interdependent relations between genres. The volume draws on a range of case studies each with a distinct genre assemblage and social agenda, exploring such areas as high stakes science, open peer review, science reproducibility, citizen science, and social media networking.

Offering new directions for future research on genre studies and digital science communication, *Genre Networks: Intersemiotic Relations in Digital Science Communication* will be of interest to scholars in these fields, as well as those working in multimodality, language and communication, and languages for academic purposes.

Carmen Pérez-Llantada is Professor of English at the University of Zaragoza, Spain. Her principal research interest lies in written genres in academic and research settings, with a focus on emerging genres in digital environments. She is the author of *Research Genres across Languages* (2021).

María-José Luzón is a Senior Lecturer (PhD) at the University of Zaragoza. Her current research focuses on the analysis of digital genres, and genre relations, for science communication and dissemination. She has published in *Applied Linguistics* and *Journal of English for Academic Purposes*, among others.

Routledge Studies in Multimodality

Edited by Kay L. O'Halloran, Curtin University

Titles include:

Multimodality and Classroom Languaging Dynamics
An Ecosocial Semiotic Perspective in Asian Contexts
Dan Shi

A Multimodal Approach to Challenging Gender Stereotypes in Children's Picture Books
Edited by A. Jesús Moya-Guijarro and Eija Ventola

Mediation and Multimodal Meaning Making in Digital Environments
Edited by Ilaria Moschini and Maria Grazia Sindoni

Multimodal Literacies Across Digital Learning Contexts
Edited by Maria Grazia Sindoni and Ilaria Moschini

Multimodality in English Language Learning
Edited by Sophia Diamantopoulou and Sigrid Ørevik

A Multimodal and Ethnographic Approach to Textbook Discourse
Germán Canale

Genre Networks: Intersemiotic Relations in Digital Science Communication
Carmen Pérez-Llantada and María-José Luzón

Southernizing Sociolinguistics
Colonialism, Racism, and Patriarchy in Language in the Global South
Edited by Bassey E. Antia and Sinfree Makoni

For more information about this series, please visit: https://www.routledge.com/Routledge-Studies-in-Multimodality/book-series/RSMM

Genre Networks
Intersemiotic Relations in Digital Science Communication

Carmen Pérez-Llantada and María-José Luzón

Routledge
Taylor & Francis Group

NEW YORK AND LONDON

First published 2023
by Routledge
605 Third Avenue, New York, NY 10158

and by Routledge
4 Park Square, Milton Park, Abingdon, Oxon, OX14 4RN

Routledge is an imprint of the Taylor & Francis Group, an informa business

Library of Congress Cataloging-in-Publication Data
A catalog record for this title has been requested

ISBN: 978-0-367-70734-7 (hbk)
ISBN: 978-0-367-70736-1 (pbk)
ISBN: 978-1-003-14773-2 (ebk)

DOI: 10.4324/9781003147732

Typeset in Sabon
by Deanta Global Publishing Services, Chennai, India

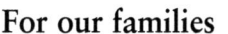
For our families

Contents

Acknowledgments

The articles of the case studies 1, 2, and 5 are licensed under a Creative Commons Attribution 4.0 International License, which permits use, sharing, adaptation, distribution, and reproduction in any medium or format. Appropriate credit to the original authors and the source has been given in each study and a link to the Creative Commons licence has been made. We reproduce original material as no changes were made. We also thank Ronald Myers, Director of Editorial *JoVE*, Cambridge, MA, for having granted us permission to reproduce some screenshots of the selected article in Case study 3.

The textual material of Case study 4 is reproduced with the permission of the researchers involved in the citizen science project "PollinatorWatch" (https://www.zooniverse.org/projects/tokehoye/pollinatorwatch), Toke Thomas Høye (Head of section for Biodiversity, Department for Bioscience, Aarhus University, Denmark) and his collaborator Isa Lykke. We also have Zooniverse's permission to reproduce this material. All the tweets/retweets in Case study 5 have been reproduced with the corresponding authors' permissions.

This book is a contribution to the project Digital Genres and Open Science, funded by the Spanish Ministry of Science and Innovation (PID2019-105655RB-I00, funded by MCIN/AEI/ 10.13039/501100011033) (http://genci.unizar.es/).

The research conducted for preparing this book has been carried out under the auspices of the Institute for Biocomputation and Physics of Complex Systems (BIFI) (http://bifi.unizar.es/) and the Instituto Universitario de Investigación en Empleo, Sociedad Digital y Sostenibilidad (IEDIS) (http://iedis.unizar.es/) at the University of Zaragoza, Spain.

1 Networks of genres in Web 2.0

This book aims to expand our current understanding of genres for science communication online through the examination of the semiotic interrelations between connected genres in media environments. It should be made clear from the very beginning that this book does not specifically seek to describe individual (or single) genres. This has already been done by the seminal literature (see, e.g. Gross and Buehl, 2016; Gross and Harmon, 2016; Kelly and Miller, 2016) and by an increasingly informative stock of current research. Rather, this book seeks to explain how specialized knowledge is constructed, transformed, reproduced, and reinterpreted within forms of genre collectivity in the digital environment. We will conceptualize these forms as genre networks in order to understand how scientific knowledge is co-constructed by different social actors (rhetors) and dialogized across expert audiences and non-specialist audiences, including the general public and diversified audiences. In doing so, and taking a theoretical stand, we also seek to foreground the timeliness of genre theory in the English for Academic Purposes genre tradition (Swales, 1990, 2004) and the value of the genre lens as an analytical tool to investigate the complex intersemiotic relations that are established within, between, and among connected genres and modes in online environments.

To set out the points of departure and with a view to proposing several theoretical and methodological considerations in closing the book, in this chapter we first offer an overview of the impact of web technologies on both the professional and public communication of science today. We also introduce the social and technological agendas that have triggered new communicative demands and prompted the emergence of new genres in response to new rhetorical exigences. In view of this, we then briefly outline the following issues: (i) how the existing repertoire of genres of science is becoming increasingly diversified; (ii) how new forms of communicating science in new media environments are supported by the main technological affordances of Web 2.0, namely, multimodality, hypertextuality, and interactivity and; (iii) from a theoretical standpoint, how evolution and innovation in genre networks involve multisemiotic interactions in science communication online. Our intention, therefore, is to provide an overview of the rapidly changing

DOI: 10.4324/9781003147732-1

social and technological dynamics of science communication through the analysis of networks of hyperlinked genres and of the intersemiotic relations within these networks. We also introduce some key terminology to help the reader better grasp the theoretical and methodological issues addressed in subsequent chapters. Finally, we provide the rationale for this book and an overview of its chapters.

Contextualizing genres online

At the outset, it is pertinent to consider some seminal theoretical conceptualizations of the term "genre" that consistently postulate a view of genres as constructs that enact social action (Miller, 1984) or social intentions (Bazerman, 1994, 1997) in human spheres of activity (Freedman and Medway, 1994). In the words of Freedman and Medway (1994: ix), there is "some kind of dialectical reflexivity between text and context, between genre and culture, between writing and subjectivity." The very fact that such dialectical reflexivity exists in web-mediated meaning-making processes makes it worth exploring how extant genres and emerging genres establish connections in complex ways on Web 2.0, the interactive web. This web is a digital infrastructure that enables the use of multiple semiotic modes while promoting a participatory culture within the global virtual community. Compared to its antecedent, Web 1.0, which was static and merely worked for content distribution, Web 2.0 provides substantially different ways of creating, sharing, and communicating knowledge through dynamic contents, multimedia elements, and interactivity tools integrated into the web architecture (Anderson, 2007).

It has traditionally been taken for granted that the main way of disseminating research widely and giving visibility to research work is through publishing articles and their related abstracts in scholarly journals. Needless to say, "publish or perish," defined in Wikipedia as "an aphorism describing the pressure to publish academic work in order to succeed in an academic career," is of great concern for every scientist. But at present, though, scientists' professional practices involve handling a much wider and, more importantly, increasingly expanding repertoire of genres. As we argue elsewhere (Pérez-Llantada, 2021b), there is *prima facie* evidence of this in today's sophisticated electronic journal platforms, where researchers can associate and have their articles hyperlinked with other texts with similar or related contents, as well as with generic innovations in web environments such as graphical abstracts, author videos, lay (or plain language) summaries, and podcast interviews, among others. Further evidence of such an unprecedented transformation in science communication practices can be found in emerging digitally mediated practices, such as blogging, microblogging, and the use of academic and social networking sites. In these new practices, the members of multidisciplinary communities can maintain informal exchanges with other peers, and disseminate and co-construct scientific knowledge in different modes—written, spoken, or both—and media to reach broad publics (Gross and Buehl, 2016; Luzón and Pérez-Llantada, 2019, 2022; Mauranen et al., 2020). In general, the shift from print media to web-mediated communication practices and greater reliance on technological affordances for communicating science online can be observed across the disciplinary spectrum, although it is certainly more noticeable in some disciplinary fields than in others (Pérez-Llantada, 2021b; Thelwall and Kousha, 2015).

Underpinning such a dramatic shift in research communication practices lie recent research and science policy agendas that operate at both global, regional—i.e. supranational—and national levels. These agendas pertain not only to the scholarly domain but also to broader socioeconomic, political, and social interests. Open Science (OS hereafter) is a major movement that stands out in these agendas. It has notably influenced researchers' activity, particularly in constructing, sharing, and disseminating their research work through drawing on the technological developments implemented in Web 2.0 (Prem et al., 2016). The OS principles stress the importance of the transparency and visibility of scientific research processes and the reproducibility of research results (Kwok, 2018). These principles also encourage researchers to publish their work in Open Access (OA hereafter). Even so, as Wynne (2006) argues, it is not clear whether OA really has made a real impact. But researchers are encouraged, more and more, to share their data online and their results with other peers to facilitate international collaboration and address social concerns over science reproducibility (Bartling and Friesike, 2014; Goodman et al., 2016; Schulson, 2018).

OS also advocates the democratization of science, aiming to promote ways of improving the accountability of science and assessing its real impact on society. This approach to scientific research has led to new forms of constructing, transforming, and co-constructing knowledge, and explains the emergence of genres that involve both researchers and broader publics, such as citizen science genres (Ball, D. 2016; Science Europe, 2017). These emergent genres target what Gross (1994) conceptualizes as public understanding of science and public engagement in science. Taking on board these socio-historical contexts and agendas, in this book we will illustrate how traditional research outreach such as journal articles and abstracts are remediated (i.e. moved from print to web environments), how they are connected with other genres online, and what processes of knowledge transformation and recontextualization are involved among them. We will also illustrate how OS and its push for OA and for the democratization of science, seen as productive ways of fostering researchers' interest in socially committed scientific research (Bonney et al., 2009; Loroño-Leturiondo and Davies, 2018; Owen et al., 2012) and enhancing society's scientific literacy, have given rise to the creation of networks of interrelated genres online. For this reason, we believe it is worth examining semiotic interrelations among and between genres in greater depth.

Genres connected online form complex multisemiotic systems for meaning making. They are supported by technological tools that have been developed to assist researchers in making their work more visible, trustable, and accessible not only to expert readers—i.e. peer scientists—but also to diversified audiences, including lay audiences. This can be seen, for instance, in the enhancement options available for research articles online. Journal platforms incorporate audiovisual innovations to enhance the contents of the articles, such as interactive maps and three-dimensional images, graphical abstracts, author videos, and audio slides. They also offer the possibility of receiving readers' feedback both in the form of online comments and responses to comments on the journal platform and in the form of views, downloads, and citations, and shares, likes, and posts in social media (Aalbersberg et al., 2012; Harmon, 2019; Pérez-Llantada, 2021b; Yang, 2017). Other examples of generic innovations that align with the spirit of the OS/OA approaches to scientific communication are open peer reviews, editors' decision letters and authors' responses to reviewers. These were traditionally "occluded evaluative genres [...] out of sight to outsiders" (Swales, 2004: 18–20) but on web environments, they are connected—and internally hyperlinked—to their associated research articles in response to the need for greater transparency of research processes (Breeze, 2019). This is also the case for other genres such as registered reports, lay summaries, video abstracts, and author videos connected with their associated articles (Cox, 2015; Mehlenbacher, 2019a; Spicer, 2014). It also applies to visualized experiment articles, also called video methods articles (Hafner, 2018), and to protocol videos (Pasquali, 2007), and data articles (Pérez-Llantada, 2021b; Shaklee, 2014), that serve to expand and enhance the contents of their related research articles. We will illustrate some of these later in this book.

Multisemiotic meaning making within genre networks online can also be traced through the characterization of emerging multimodal genres (Kelly and Miller, 2016), as we will also illustrate in this book. This is based upon technological affordances—mainly hypertextuality and interactivity—that are used in both peer-to-peer informal communication and public communication of science. By way of illustration, academic and social networking sites (ASNSs) today represent virtual spaces for informal knowledge exchange and knowledge co-construction among the members of the scientific community. Other emerging genres combine different semiotic modes to create meanings and allow wide scientific knowledge dissemination among broad publics. Blogging and microblogging practices are a case in point here (Cody et al., 2015; Luzón, 2013; Mauranen, 2013; Puschmann, 2014; Shema et al., 2012). Lastly, intersemiotic relations are created within genres that have emerged for public engagement in science, such as citizen science projects. These projects are supported by visually appealing web portals with multimodal, hypertextual, and interactivity functionalities that enable user content generation and dialogic exchange between scientists and citizens using a single electronic platform. In this case, such intersemiotic relations are established within the genre itself, which acts as a macro genre containing embedded genres. We will see an example in one of the case studies. In the following sections we briefly introduce other key theorizations that we also find useful for describing and interpreting intersemiotic relations in connected genres online.

Knowledge transformation processes

Moving beyond the analysis of single genres and exploring networks of interrelated genres necessarily raises the issue of how knowledge is actually transformed and recontextualized when it is taken up from one genre and used in another genre. Over the past decades, there has been considerable scholarly interest in the study of knowledge transformation in processes that involve the recontextualization of scientific knowledge to express different viewpoints and perspectives and to adapt scientific knowledge to different audiences. "Recontextualization" is, therefore, an important concept required to understand discoursal uptake within networks of genres online. "Recontextualization" has been defined as a complex and dynamic transformation of meaning from one context to another, which involves adaptation to the practices of the new context in order to meet the expectations and needs of the audience (Blommaert, 2005; Linell, 1998). The term has also been applied to characterize science popularization genres and has been taken on board in the field of Languages for Specific Purposes and English for Academic Purposes (LSP and EAP hereafter), professional communication, and discourse studies on emerging genres (Luzón, 2013; Orpin, 2019). In this book, this concept is particularly important insofar as we postulate that every genre network incorporates multiple views and a polyphony of voices addressing the same phenomenon. At times, these views represent the perspectives of the same social actors (or rhetors) who may adopt different identities. At other times, the networks contain views and perspectives from different social actors. In either of those cases, the contents are textualized differently when they move from one genre to another genre because each genre fulfills different communicative purposes and targets specific audiences.

In addition to recontextualization, other conceptualizations have also been proposed to better understand knowledge transformation processes and discursive uptake. These are "remediation," "(re-)entextualization," "genre hybridization and embedding," and "resemiotization." We briefly summarize them below.

Remediation

Remediation is defined as "the formal logic by which new media refashion prior media forms," for instance photography remediates painting and television remediates film (Bolter and Grusin, 1999: 273). According to Bolter and Grusin (1999) media do not operate in isolation but always reproduce other media, adopting and adapting their techniques and forms. New digital media transform and improve earlier media—e.g. television, books—while retaining some of their features. The digital medium has specific potentials and constraints that determine how the message is produced and distributed. When pre-digital genres migrate to the web, they are usually transformed and enhanced by means of the possibilities afforded by Web 2.0 infrastructures—as stated earlier, hypertextuality, multimodality, and interactivity. An illustrative example is provided by Andersen and Van Leeuwen (2017) in their analysis of the UK version of the fashion shopping site Zalando (http://www.zalando.co.uk). The shopping site remediates shopping service encounters and enables the user to perform similar actions to what they do in the shop. Online service encounters reconstruct face-to-face shopping by drawing on the semiotic resources available in the digital medium—e.g. written language, photography, layout—which replace the spoken language and gestures used in face-to-face interactions.

Remediation may involve genre change leading to genre evolution and often to the emergence of new genres. As explained by Østergaard and Bundgaard (2015: 124), "genres emerge as amendments, accommodations or suitable modifications of already existing text types with a view to provide an adequate discursive response to a novel kind of situation (*or with a view to exploit the affordances of new technology*)" (2015: 124) (our emphasis). An example of how remediation has transformed genres is the so-called "enhanced article" (Harmon, 2019) that, as we mentioned previously, consists of a journal article online accompanied by linkable resources—e.g. datasets and supplementary materials—and hyperlinked or embedded genres such as lay summaries, graphical abstracts, or author videos, among others. These accompanying genres have been conceptualized as extensions of genres or add-on genres (Aalbersberg et al., 2012; Harmon, 2019). Although the genre remains very much the same (Anderson, 2009; Pérez-Llantada, 2013), online innovations enhance the article contents in several ways. For instance, interactive maps are used to enrich the article content and enhance the visualization of information (Aalbersberg et al., 2012), and video abstracts and lay summaries are used to "increase the visibility, impact, and transparency of scientific research" while creating "direct pathways between scientists and the general public" (Kuehne and Olden, 2015: 3585).

Re-entextualization

Another concept used to study knowledge recontextualization processes is the concept of (re-)entextualization. Gimenez et al. (2020) propose this term to describe the processes by which scientific knowledge begets text—or is entextualized—in a text and, subsequently, re-entextualized in other texts that may or may not necessarily belong to the same situational and socio-historical contexts. As these authors explain it, "written and spoken texts are recycled in a variety of ways and across a range of contexts" (Gimenez et al., 2020: 294) and therefore reach different audiences, from scientists or readers interested in popular science to broader publics with diverse interests. To examine trajectories of texts on science-related issues these authors borrow Blommaert's (2005) concept of "re-entextualization" to track processes of knowledge transformation manifested in the trajectory of a scientific text that involve "re-interpretation, re-organization, and re-focusing" (Gimenez et al, 2020: 297). These authors contend that the texts undergo "transformational processes that involve issues of social power, authority and access that require new analytical tools to surface more clearly." Aligning with these authors' claims, in this book we argue that explorations of connected genre exemplars—and not just necessarily sequential text trajectories—need to put the focus not simply on the strategies or ways of simplifying, expanding, or reformulating contents to reach the interests of the Internet publics. In our view, these explorations need to examine critically the social actors that have produced these texts, the way the actors' individual narratives construct an identity—at times, even several identities—and the rationales or motivations that make these different actors, as rhetors, transform the knowledge textualized in previous texts in particular ways. Accordingly, the analytical approaches and frameworks that we will propose in Chapter 3 will aim to assess critically the complexities underpinning the interrelations of genres, modes, and media when dealing with genres that work in conjunction with other genres in the digital medium. For Gimenez et al. (2020: 295) "re-entextualization" as an analytical tool effectively applies to processes by which "knowledge is re-interpreted, changed, and even lost as a result of issues of power, authority and access implicated in such processes."

At this point, it is also worth recalling earlier studies that enquired into the issue of "science accommodation" (Fahnestock, 1986), a concept defined to explain the different ways through which rhetors adjust the conceptual depth of scientific knowledge to the values, assumptions, and needs of diversified audiences. Taking the case of science popularization genres, Bondi et al. (2015) explain that in order to adjust expert knowledge scientists draw upon strategies of reformulation and paraphrasing that make specialized contents accessible and relevant to broader audiences (see also Gotti, 2014). These studies emphasize, however, that recontextualization does not involve only making knowledge accessible; it also means adapting it to fit the new context, to meet the expectations of audiences, and to achieve the purposes of the genre, as we aim to illustrate with the case study research.

Hybridization and embedding

The concept of hybridization can help us clarify the outcome of discoursal uptake and recontextualization processes in digital communication as well as the intersemiotic relations created between connected genres. This concept may be useful when the network contains genres hyperlinked to other genres from which they borrow linguistic and/or rhetorical features. Early, but very influential and timely, work by Jamieson and Campbell (1982) conceptualizes them as "rhetorical hybrids" as they are new genres that mix generic features and appropriate linguistic features of existing genres or antecedent genres. This definition has been adopted in recent scholarly work on digital genres to explain the hybrid nature of genres such as blogs (Herring et al., 2004) and websites (Catenaccio, 2012). The term "para-scientific genre" is also used to characterize emerging genres online that borrow authority and expertise from genres associated with professional, expert-to-expert science communication (Kelly and Miller, 2016). On the other hand, the scholarly literature has also foregrounded the hybrid nature of specialized discourses. For instance, Bhatia (2004: 87) discusses features of interdiscursivity in professional communication that result from the appropriation "of generic resources from a specific genre for the construction of another," eventually merging different discourse types in a single genre. Motta-Roth and Scherer (2016) also offer an insightful discussion of interdiscursivity features to refer to the merging of discourses of science, pedagogy, and journalism in science popularization genres online.

The concept of genre embedding is also worth highlighting here, as this generic phenomenon can also be observed in networks of genres online. Genre embedding, which involves incorporating a text representing a genre within the matrix of another genre, is useful for understanding multisemiotic meaning making in connected genres online (Kwásnik and Crowston, 2005; Luzón, 2017; Orlikowski and Yates, 1994). Again, the retail site of Zalando clearly illustrates the role of embedding in the construction of digital genres. The site is defined by Andersen and Van Leeuwen (2017) as a "multi-generic structure potential," which, thanks to the variety of genres it includes, enables the text producer to achieve various communicative purposes, such as informing, persuading, or selling, *inter alia*. The site embeds several "micro genres" (Orientation, Catalogue, Product Information, Purchase, Fashion Magazine, Lifestyle Magazine, Street Style), which readers can combine in different ways to create macro genres. These macro genres are constructed by the website user by navigating between and through some of these micro genres to achieve their goals, such as getting information or buying. We will analyze an analogous phenomenon in genre networks related to public engagement in science, specifically in citizen science projects, that instantiates, as we will see in one of the case studies, the multigeneric structure potential of macro genres to achieve diverse communicative purposes—namely, informing, persuading, and instructing—by embedding several verbal and audiovisual micro genres.

Resemiotization

Much work on recontextualization strategies has focused on linguistic strategies such as reformulation or exemplification that involve the verbal mode. However, more recent research has explored cross-modal recontextualization, or "resemiotization," that is the process of reproducing meaning across different modes and modalities (Iedema, 2003; Leppänen and Kytola, 2017).

The movement of pre-digital genres to the digital media means that genres can also evolve by harnessing the multimodal affordances of these media. Digital media offer the possibility of combining a multiplicity of modes—verbal, aural, visual—to achieve the purposes of the genre. The concept of "multimodality" comes from new media and communication studies and studies on genre and discourse (Bateman, 2008; Engberg and Maier, 2015; Jewitt et al., 2016; Prior, 2013; Salmose and Elleström, 2020). Given that our point of departure in this book is the view of genres as social action, we adopt Kress's (2010) social-semiotic perspective on multimodality, a key feature of digital genres and a crucial component of web-based forms of genre collectivity that are inherently multimodal. Domingo et al. (2015: 252) define mode as "a set of resources, socially made, to enable us to deal with social needs and achieve social purposes." Examples of modes are still images, moving images, written language, gestures, spoken language, music, and non-verbal sound (Tseronis and Forceville, 2017). When several modes are combined in a communicative event, they form a multimodal ensemble, where all the modes contribute to meaning making. Each mode has specific modal affordances (Kress, 2010), for example their potentialities and constraints or what they can express better than other modes, and therefore the modes forming part of a multimodal ensemble contribute in different ways to meaning making (Jewitt, 2016). Engaging in multimodal genre network analysis involves analyzing genres by examining both linguistic and non-linguistic modes or modalities in genres—e.g. sound, static images, moving images—to take on board all the semiotic representations of meaning making that are "socially situated" (Bhatia, 2008; Paltridge, 2012: 170). Complementing the genre perspective, the multimodal perspective sheds further light on the way social actors—i.e. the rhetors—achieve various social and communicative purposes that range from informing, explaining, persuading, or counterclaiming, to entertaining or educating in issues of science by integrating multisemiotic resources (Mauranen, 2013; Scotto di Carlo, 2014). Resemiotization is, therefore, a very valuable concept to describe how language, operationalized in semiotic modes and codes, "ties back into antecedent language at the same time that it anticipates subsequent language" (Scollon, 2008: 233).

The effects of context collapse

In defining genre as a type of communicative event, Swales (1990: 58) explains that "exemplars of a genre exhibit various patterns of similarity in terms of structure, style, content and intended audience" and that it is precisely these patterns of similarity that make genres prototypical and recognizable by their users. In fact, "intended audience" is, along with "communicative purpose," one of the defining criteria for genre identification (Swales, 1990). That said, the traditional concept of intended audience as conceptualized in genre and rhetorical genre studies becomes problematic when analyzing genres in the digital medium because science communication online makes it difficult to establish clear-cut distinctions between "intended audiences" or virtual communities. Such problematization can be clarified to some extent if we take on board the concept of "context collapse" used to refer to how social media flatten "multiple audiences into a single context" (Brandtzaeg and Lüders, 2018: 1). Genres in online environments are open to audiences in a variety of spheres (Boyd, 2002; Marwick and Boyd, 2011; Trench, 2008) and with diverse degrees of expertise. Therefore, it is difficult to say that a genre exemplar online specifically targets a single, well-defined intended audience. For example, when one seeks to identify who the audience of an enhanced publication is, one could hypothesize that the article online is primarily associated with a discourse community which "has a threshold level of members with a suitable degree of relevant content and discoursal expertise" (Swales, 1990: 27), namely, the scientific community. This is a readership specialized in the field and thus able to interpret the article contents. In other words, neither the conceptual depth, or degree of "densification" (Leech et al., 2009) of discourse, nor the grammatically compressed style that characterizes academic writing (Biber et al., 1999) are a threshold, metaphorically speaking, that the reader needs to cross to fully understand the article contents. In contrast, a readership with little or no background in the content field might find such conceptual depth and lexically dense, non-explicit discourse style thresholds difficult to penetrate, and might then opt for accessing contents of texts that are less conceptually dense, for example, a lay summary of the article intended to disseminate article contents beyond expert audiences.

We can think of other examples that illustrate the problematization of the concept of intended audience. For instance, we can consider readers' information access behavior in online environments. While traditional print genres involved a linear reading process, on the web the materiality of the digital medium and, in particular, the technological possibilities for internet navigation—i.e. text modularity, hyperlinking, and interactivity—offer readers different entry points for accessing knowledge, perspectives, and views of scientific contents (Miller and Fahnestock, 2013). Each user may choose their preferred entry point to access knowledge and select contents with different degrees of specialization or conceptual depth, and also with different viewpoints and perspectives towards such contents (Casper, 2016; Pérez-Llantada, 2013). Internet modularity, hypertextuality, and interactivity are therefore features worth exploring in connected genres online so as to understand the role they play in framing intersemiotic meaning making in genre networks online.

Other pertinent arguments about the challenges of addressing broader audiences in collapsing contexts have been made by rhetorical genre studies and studies of social-semiotic theory. These studies explain that in online communication arguments need to be very explicit to maximize comprehensibility and interpretability (Buehl, 2016). Cheryl Ball (2016: 56) remarks that new genres "tend to be infrastructurally complex" and therefore arguments are even more explicit and compelling when the text relies on multimodal rhetoric or the combination of verbal and visual rhetoric (Mehlenbacher, 2019b; Roque, 2017). Accompanying textual/verbal rhetoric, the use of visual rhetoric has also been extensively discussed when examining the impact of images in science communication online (Kress and Van Leeuwen, 2001; Pauwels, 2006). For example, Sancho-Guinda (2019) uses the term "promoemotional science" to describe the graphical abstracts that accompany the verbal abstracts of research articles online. The persuasive function of visual rhetoric in the context of knowledge transformation processes has also been identified as a key feature of research group blogs (Luzón, 2020) and crowdfunding medical campaign narratives (Paulus and Roberts, 2018). Paulus and Roberts specifically note that crowdfunding projects build arguments around the affective conditions of the person in need of medical assistance, which is enhanced by means of visuals. The use of visual rhetoric involves considering the cultural assumptions and different cultural expectations of diversified audiences to produce positive impressions and prompt trust in science. In genre networks online, photographs, still and moving images, and videos are used to create rhetorical effects that range from informing and instructing to persuasively appealing to the imagined audiences.

In discussing the role of rhetoric in public communication of science, Gross (1994: 6) explicitly mentions that one problem is the assumption that the target audience is not knowledgeable about science, while this is not necessarily the case. As we also noted elsewhere (Pérez-Llantada, 2021b), Gross's distinction between two different models, the "deficit model" and the "contextual model," makes this point clear. The deficit model, defined as solely cognitive and asymmetrical, "depicts communication as a one-way flow from science to its publics" and a type of communication in which researchers do not need to build trust because the public is persuaded of the value of science" (Gross, 1994: 6). On the other hand, the contextual model, symmetrical and dialogic, "depicts communication as a two-way flow between science and its publics." Along similar lines, in his examination of science popularization genres Myers (2003: 265) brings to the fore the problematic nature of the concept of intended audiences. Myers argues that these genres should not be viewed "as a one-way process of simplification, one in which scientific articles are the originals of knowledge that is then debased by translation for a public that is ignorant of such matters." In our view, the contextual model is more appropriate than the deficit model to understand intersemiotic systems and meaning-making processes in connected genre networks online and to illustrate the effects of context collapse in new media environments.

Trench (2008: 185) is also very explicit as regards the "erosion of boundaries" on the Internet, claiming that the boundaries between professional and public communication are more porous than ever, "facilitating public access to previously private spaces," and thus "turning science communication inside-out." We can also find in the literature judicious arguments that have important implications for understanding intersemiotic relations among and between connected genres online. One of them is Burns, O'Connor, and Stocklmayer's (2003: 191) view of Internet audiences as multiple and diverse audiences with different interests in science and varying commitments to scientific activity. These authors sensibly conclude that effective science communication online relies on the appropriate use of "skills, media, activities and dialogue" through which scientists can produce a range of responses using, as these authors put it, the AEIOU analogy, "Awareness, Enjoyment, Interest, Opinion-forming and Understanding." The case studies we analyze in this book illustrate the diverse audiences that genres working in conjunction with other genres can reach and the range of responses they can elicit from them.

Aim of the book

This book aims to bridge the theory of science communication with the real-life practice of communicating scientific knowledge in new media environments by drawing on genre analysis as both a theory and an analytical approach. Essentially, the book aims to portray how genres connected to other genres in digital environments enact complex social actions. The book also aims to describe the range of communicative intentions underlying processes of knowledge (re-)entextualization across texts and how these intentions are textualized—"entextualized" to use Gimenez et al.'s (2020) term—by rhetors in particular ways in different genres. In taking up entextualized discourse, a rhetor or rhetors reorganize(s), reinterpret(s), and refocus(es) them to achieve their communicative intentions and satisfy the expectations of their imagined audiences. Investigating genre interrelations in science communication online is a topic of prime importance and, in our view, very timely while increasingly diverse social interaction practices on issues of science rely to a great extent on the technological possibilities offered by Web 2.0. To advance the existing theoretical understanding of Science 2.0, we use the genre analytical lens to examine the ways in which science is communicated in the digital environment. But rather than characterizing single text types or genres we look at the intersemiotic relations established among the genres forming a given network.

Our main theoretical and interpretative framework is (critical) genre analysis (Bhatia, 1993, 2004; Swales, 1990, 1993). We apply this framework to explore semiotic interrelations. Following Herring (2019), we understand that online communication is fundamentally multimodal and for this reason we do not distinguish between monomodal and multimodal genres. To explore the interrelations established in multisemiotic meaning-making systems such as genre networks online, we propose the concept of "genre network" that we broadly define as a kind of genre collectivity or set of connected genres online. This concept is not dissimilar to the types of "genre assemblages" in professional communication settings described by Spinuzzi (2004), namely, genre sets, genre systems, genre repertoires, and genre ecologies. We prefer to use the concept of genre network, instead of other terms used in the seminal literature such as genre system, or genre set, that are fairly similar conceptualizations, because for the purposes of this book the word "network" itself already reinforces the idea of interconnectedness or interrelatedness among genres that we want to stress and enquire into. Interconnectedness will thus be a quality shared by the networks of genres that we selected as case studies for illustration purposes.

We contend that in a digital environment genres should be viewed not only as stand-alone constructs with their own generic integrity, but also as open constructs associated with other genres and, hence, sharing features with them at the semantic, syntactic, discoursal, and/or rhetorical level, as we explain in further detail in Chapter 2. Essentially, we claim that connected genres online should necessarily be viewed as interrelating constructs, closely linked to other genres, modes, modalities, and other semiotic forms of meaning representation. Looking at networks enables us to capture the full picture of such intersemiotic relations. The types of interrelationships established between or among genres within a network will then be our focus. To the best of our knowledge, this analytical approach has not yet been undertaken to date in relation to professional and public science communication on the Internet. We therefore expect to move the field ahead by offering descriptions of connected genres and interpreting their interrelations with other genres. Like others, we draw on the assumption that the very diverse audiences of the Internet may choose different entry points to access information on issues of science that is textualized with different levels of conceptual depth and that conveys different views and perspectives towards scientific contents (Engberg and Maier, 2015; Maier and Engberg, 2019). This assumption invites us to enquire into the genre network as a whole and not into single genres. To validate this assumption, we analyze five case studies that involve multisemiotic modes of expression and illustrate different intersemiotic relations. This case study research purports to offer a more comprehensive understanding of digital communication models and to give further insight into how social actors—and science stakeholders, including citizens—communicate issues of science on the web. With the help of the case studies we also aim to highlight the "wide functional heterogeneity" of genres in the Internet (Giltrow and Stein, 2009: 10).

More broadly, in this book we aim to offer evidence to support the view of genres as constructs that are characterized by their openness and their dynamism, or their capacity to adapt, mix, and hybridize with other genres in response to new communicative demands and social exigences. With some exceptions, research on digital genres to date, though very prolific and insightful, has only provided descriptive characterizations of single genres, as we noted earlier. For this reason, we believe it is important to move forward and engage in the analysis of genre interactions in the digital medium to understand the dynamics of multisemiotic meaning making online. We also expect that the analytical approach taken in this book sets the grounds for developing in future research a more comprehensive, in-depth view of multisemiotic forms of representation in genre networks and their functional heterogeneity to enact social actions. In our view, this analytical approach, still exploratory, allows us to better understand more accurately how science is communicated, dialogized, and co-constructed in Web 2.0 today.

Methodologically, we also advocate the complementariness of genre analytical methods with other analytical procedures to capture the complete picture of intersemiotic relations online. Here, we align with Bhatia's (2004: 156) claim that different analytical procedures need to be integrated to fully understand "genres in the real world of discourse." We devote Chapter 3 to discussing such procedures in more detail. Integrating these procedures in the exploration of connected genres allows us to describe in a more precise way what Bhatia (2004: 156) defines as text-internal features of genres—i.e. "rhetorical moves, discourse strategies, regularities of organization, intertextuality and some aspects of interdiscursivity," and text-external factors such as "participant relationships, and their contributions to the process of genre construction, interpretation, use and exploitation in the context of disciplinary, professional and other institutional practices and constraints." We believe that this integral approach is crucial to understanding intersemiotic relations because the presence or absence of certain text-internal features are intrinsically associated with the text-external factors that shape and constrain the way scientific knowledge is entextualized and subsequently re-entextualized across genres and modes in particular ways. Finally, informed by these theoretical and analytical approaches we postulate research-informed pedagogies of digital multimodal composing in order to raise awareness of multisemiotic meaning making and processes of knowledge transformation—e.g. recontextualization, remediation, re-entextualization, and resemiotization. As practitioners in the field of Languages for Academic Purposes (LAP hereafter), we also propose possible ways of guiding and instructing scientists in effectively communicating and understanding science online.

Overview of chapters

Having provided the rationale and scope of the book in this introductory chapter, in Chapter 2 we summarize the interpretative theories and constructs that we apply to make sense of the intersemiotic relations among genres online and understand genre connections in relation to the situational and socio-historical contexts that shape them. The theories and concepts that we will adopt belong to multidisciplinary fields, ranging from literary criticism, rhetoric and composition studies, genre and discourse analysis, academic writing, and professional communication to social theories, new media studies, and computer-mediated communication. All these fields have a common interest in understanding the contexts that affect multimodal text composing and are relevant to multisemiotic meaning-making processes, in this case, in the virtual or digital space.

Having set the theoretical grounds, in Chapter 3 we explain how we apply our proposed analytical framework. Essentially, we complement genre analytical procedures with data from corpus-assisted analytical methods, and we also carry out intertextual, discourse, and multimodal analyses. As shown in the case studies, the complementarities of these procedures enable the analyst to trace shared semantic meanings, concurrent texts with different conceptual depth, as well as intertextual and interdiscursive features, which serves to better understand genre interrelations in new media environments. As discussed in this chapter, we combine the genre lens with other multiple perspectives to explore how modes of meaning making for science communication operate in Web 2.0. We explain how the genre lens enables us to enquire into aspects of genres and textuality, genres and intertextuality, and genres and multimodality. We purport to showcase that in putting together the outcome of this inquiry the analyst is in a better position to provide a more accurate account of the multisemiotic forms of representation online.

In order to illustrate this analytical approach, Chapter 4 includes five case studies in which we apply different analytical possibilities for examining online genre networks. The case studies cover exemplars of genres of both professional and public communication of science and, therefore, texts produced by both scientists and other science stakeholders. For the sake of diversity, each case study centers on a distinct genre network with a different social and/or rhetorical exigence. Case study 1 focuses on the analysis of a network whose core genre is an expert-to-expert genre—a journal article published in *Nature*—that is intertextually linked to genres for public communication of science. In Case study 2 we analyze a sequential chain of genres retrieved from the journal *eLife*, namely a journal article, its associated reviewers' reports, the authors' response to the reviewers and the authors' impact statement and digest (or plain language summary), all posted on the journal website. Case study 3 exemplifies a multisemiotic genre set used for reporting scientific methods, and not reporting of results, unlike Case study 1 and Case study 2. In this case study we examine meaning making across semiotic modes in a video methods article from the *Journal of Visualized Experiments* and its associated written article. Along with the professional communication of science, addressed in the first three cases, we also showcase intersemiotic relations in genre networks for public communication of science. In Case study 4 we analyze a genre network representative of public communication of science online, namely a citizen science project from the Experiment.com web portal. In this case we specifically enquire into aspects of text modularity, genre hybridization, and genre embeddedness. Case study 5 also illustrates a genre network of public communication of science. The focus of analysis, though, is centered on the multimodal and semiotic resources used in the process of taking up the content of a research article published in *Nature Communications* and re-entextualizing it in online popularization genres and tweets. These cases will allow us to clarify how semantic meanings taken from a specific context are recontextualized and repurposed to make them suitable for a new context for addressing the information needs of non-specialist, diversified audiences, and for the purposes of the rhetors that create the texts.

The selected cases form part of small-scale specialized corpora compiled for the project Digital Genres and Open Science, an ongoing national-based Spanish research project that seeks to address the problematization of the traditional concept of genre and examine genre interrelations when communicating science on the Internet.[1] These corpora are representative of genres for both the professional and public communication of science online, from journal articles, short summaries, graphical abstracts, peer reviews, and lay summaries, to citizen science project websites, blog posts, and tweets, among others. Moreover, because we also aim to contribute to building a better understanding of the knowledge base relating to some of the goals set in the United Nations 2030 Agenda for Sustainable Development, our data sources consist of closely linked science or science-related texts online whose themes are associated with two of these Sustainable Development Goals (SDGs): SDG 3. *Good health and well-being* and SDG 13. *Climate action*. Table 1.1 summarizes the selected studies upon which we base our enquiry into genre interrelations.

There are several reasons why we selected case study research to illustrate our theoretical and methodological claims. The first one, as Flowerdew

Table 1.1 Overview of data sources for case studies

Case study	Thematic focus	Description
Case study #1	SDG 13. *Climate action*	An enhanced publication of a remediated genre that is hyperlinked to genres for public understanding of science
Case study #2	SDG 3. *Good health and well-being*	A sequential chain of research genres related to science gatekeeping processes and add-on genres for article enhancement
Case study #3	SDG 3. *Good health and well-being*	A genre set formed by a video methods article and its associated written methods article
Case study #4	SDG 13. *Climate action*	A citizen science project for collaborative research and its embedded micro genres
Case study #5	SDG 13. *Climate action*	A research article and hyperlinked online news report/article and associated tweets and retweets

(2012: 179) explains, is that case studies make it possible to write a close analysis of the texts, "an approach underpinned by Fairclough's concept of CDA" and also applied in (critical) genre analysis of specialized discourses (Bhatia, 2004). Our analysis focuses on the specific details of a text that reveal semiotic interrelations with other texts in each genre network. The second one is that using cases it is also possible to enquire into different layers of the texts or levels of textual analysis, as we explain in Chapter 3. Such enquiry can be informed by corpus data such as frequency and keyword lists, keyness, wordgrams, or measures of syntactic and lexical complexity, among other possibilities, to better understand the entextualization of scientific knowledge in terms of language, discourse style, and register features and in relation to features of authorial positioning, stance, and evaluation. In addition, case studies are of sufficiently manageable size to be able to adopt an in-depth qualitative approach to the texts, for example when applying taxonomies of lexical, grammatical, and discourse features, or identifying move boundaries or specific segments of texts. This analytical procedure also seems feasible for describing multisemiotic meaning making across connected genres online. Corpus and discourse tools can also inform meaning-making processes in and across the texts, issues of self-representation and identity construction, and the worldviews and underlying ideologies of the rhetors that create the different texts. Even tacit meanings and textual silences can be traced through automatized corpus tools if appropriate cut-off points are established.

Finally, Chapter 5 wraps up the main results of the case study research to assess critically how the analysis of connected genres can be moved forward, both theoretically and methodologically. From a theoretical standpoint we discuss the implications of the case study findings in relation to "the simultaneity, the inseparability of form, meaning, and action, of individual, social, and cultural context, of actual genres and genre-ness" (Devitt, 2009: 46). As further discussed in this closing chapter, Devitt's concept of "inter-genre-ality," or the extent to which "genres take up forms from the genres with which they inter-act," in this case in the digital space, paves the way for suggesting several pedagogical orientations. In light of these theoretical and analytical considerations, we advocate learners' rich input and exposure to multiple audience(s) writing practice to help them become effective communicators of science in and for society. We also propose some pedagogical applications for teaching and learning academic and digital literacy, including multimodal skills composing.

Book audiences

Our point of departure is, first and foremost, genre and LAP, discourse and multimodality, and the book will be of interest to researchers in these fields. We also hope it will appeal to researchers in multidisciplinary fields, such as science communication, writing research, new media studies, and social media studies. The book may also be of interest to researchers investigating activity theory in organizations (Orlikowski and Yates, 1994) and those specializing in complexity theory and dynamic human-centered communication systems theory (Lang, 2014). In fact, the view of natural language as a complex adaptive system proposed by this perspective can be set in parallel with the dynamic interactions supported by the interrelated genres online and the new communication practices that make genres adapt, evolve, and emerge in accord with the current socio-historical context.

Note

1 "Digital genres and Open Science: An analysis of processes of generic hybridity, innovation and interdiscursivity" funded by the Spanish Ministry of Science and Innovation and the Agencia Estatal de Investigación (project code PID2019-105655RB-I00, MICIN/AEI/ 10.13039/501100011033).

2 Genres and intersemiotic relations

In rhetorical genre studies, genres are described as being constituted by substance (semantics), form (syntax), and rhetorical action (pragmatics) (Bazerman, 1994, 2004b; Miller, 1984). These features enable genres to accomplish their intended communicative goals or, to put it differently, provide responses to the rhetorical exigences of the specific situational and socio-historical contexts in which they are used. This means that in order to describe and interpret how multisemiotic modes are orchestrated in genre networks online, and how intersemiotic relations among genres and modes are established, we need to bear in mind how such contexts impose constraints on the form and substance of genres. A clear example of contexts that have imposed constraints on genres is the existing socioeconomic and research policy agendas in research publishing, namely Open Science and Open Access, mentioned in the previous chapter. These agendas, for example, have triggered the emergence of genres that aim to give greater visibility to research processes and to communicate science beyond specialist audiences (Bartling and Friesike, 2014; Fecher and Friesike, 2014; Goodman et al., 2016; Kwok, 2018). Because we have already addressed these agendas in detail in previous works (Luzón and Pérez-Llantada, 2019, 2022; Pérez-Llantada, 2021b), we will not expand on them any further here, but we would like to stress once again that both agendas keep on shaping (and constraining) the nature of genres and the way genres establish relations with other genres in new media environments. We will see some examples in the case study research.

To better situate the inherent dynamics of genre networks online and give readers a more detailed theoretical rationale for analyzing them, in the following sections we briefly explain our view of "genre," highlighting those concepts that are salient in our analysis of the dynamics of networks of genres online. We also review current genre studies that have addressed the technological transformation of science communication. In addition, we outline the main analytical framework that we will apply to investigate intersemiotic relations between and among genres and modes. Finally, we sketch out the case study research we conducted to address aspects of language semiotics and discuss critically the intersemiosis of coexisting modes of meaning making.

DOI: 10.4324/9781003147732-2

Genre openness

In the genre and LSP tradition, genres have been characterized using two key criteria: intended audience and communicative purpose(s) (Askehave and Swales, 2001; Bhatia, 1993, 2004; Swales, 1990, 2004). As we have argued elsewhere (Pérez-Llantada, 2021b), applying the former criterion, intended audience, has become problematic when investigating genres in the digital environment, as genres are subject to the effects of the collapsing of contexts in Internet communication. The latter criterion, communicative purpose(s), remains valid to date as long as the analyst takes on board the claim that each genre does not perform a single communicative purpose but rather, sets of communicative purposes (Askehave and Swales, 2001) and private intentions, what Bhatia (2004: 73) calls "hidden agendas." We will see in Chapter 4 how assigning a single communicative purpose to a genre is more complex than it may appear at first sight, and even more complicated when we aim to explore the intersemiotic relations established among the genres constituting a network online. On the other hand, rhetorical genre studies, which conceptualize genres in terms of substance and form, aim to understand the way in which both meanings and formal features of genres are shaped and constrained by their social exigences (Bazerman, 1994, 2004; Miller, 1984). In our view, this approach to analyzing genres remains valid for investigating genre connections in digital science communication. In fact, recent work on genres in media environments has applied it concurrently with the concept of the erosion of boundaries suggested by Trench (2008). This latter concept has, for example, been the basis for characterizing para-scientific communication in recent work (Miller and Kelly, 2017).

Studies on LAP and professional communication have also developed several useful concepts for describing the nature of genres and, more importantly considering the scope and goals of this book, the individual integrity of each genre in relation to "the possibility of appropriation of generic resources to create new forms" (Bhatia, 2004: 112). Because we view genres as open categories of discourse, the concept of generic integrity is also timely for examining genre interrelations and multisemiotic meaning making in science communication on the Internet. Equally valid, in our view, are the concepts of genre mixing or blending, that involve the appropriation of "generic resources from a specific genre for the construction of another" (Bhatia, 2004: 87); the concept of interdiscursivity, or the "appropriation of language and discursive resources" associated with other genres (Bhatia, 2004: 90); and the "colonization" of a genre by another genre, for example, the colonization of genres associated with academic discourse by promotional discourses associated with the field of advertising. In our view, these textual phenomena enable the analyst to determine the individually recognizable integrity (Bhatia, 2004: 87) of the genres forming the network and, at the same time, the openness of genres (i.e. their capacity to connect to other genres), in order to better understand the inherent meaning-making dynamics of genre networks.

We also think it is important to adopt other concepts developed in the genre literature that appear to cut across the above-mentioned genre traditions and can help to clarify processes of knowledge transformation and recontextualization across genres in new media environments. For example, the concept of "intermediary genre," defined by Tachino (2012: 456) as "a genre that facilitates the uptake of a genre by another genre," can be used to understand how two or more "otherwise unconnected genres" are connected "to make uptake possible." Drawing on Tachino's (2012) conceptual development of intermediary genres and Freadman's (2002) concept of uptake, Bray (2019: 156) provides the example of "how online news genres take up a scientific research article about climate change over a period of one year" to explain how the authors of online news genres—i.e. news reports and editorials—use the press release, the short texts accompanying articles online, and the article abstract, to mention a few, as intermediary genres that take up the claims from the research article. Thus, the concept of intermediary genre can better inform the way in which knowledge claims expressed in a scientific article are reformulated, simplified, adapted, expanded and, more generally speaking, transformed into, say, online news, lay summaries, and blogs, all of them falling outside the domain of expert-to-expert communication of science. In our view, when analyzing genre networks, understanding the role of intermediary genres within the network can help us elucidate how scientists communicate science with diversified audiences and how scientists and/or other social actors, as rhetors, take up knowledge claims from previous texts and transform them to compose other genres. These can be as diverse as tweets on the contents of an article, or a blog post written by the author of the article or by a science journalist. Some of the cases discussed in Chapter 4 illustrate this point.

The concepts of uptake and recontextualization (Freadman, 2002; Linell, 1998; Luzón, 2013; Smart and Falconer, 2019) are fundamental in the analysis of intersemiotic relations in and among genres in online environments. Specifically, as explained in Smart and Falconer (2019: 202), they serve to clarify how the contents conveyed in a given genre are reinterpreted and repurposed to achieve other rhetorical goals. These contents, then, may differ to a greater or lesser extent from those of the text (i.e. the source text or pre-text) the contents have been taken up from. For example, by tracking intertextual chains of discoursal uptake and (re-)entextualization in one of the case studies it is possible to see how the contents and claims of a research article are taken up in the reviewers' reports and also re-entextualized by the article authors themselves to compose two post-texts, an impact statement and a lay (plain language) summary, or digest. Both are promotional genres for article enhancement and wider dissemination of the article contents. This first-hand example further illustrates the inherent complexity of interrelated genres in online environments, which invites the examination of genre connections at various textual levels (or layers of the texts, as we will call them), namely semantics, grammar and syntax, and discourse pragmatics. The generic interrelations within a given network of genres can constitute an even more complex intertextual chain of uptake and recontextualization if several modes are involved, for example when the authors of a journal article recontextualize their knowledge claims in multimodal genres—e.g. a video abstract, an audio slide presentation, or a podcast interview—by this means contributing the enhancement of the article contents. Another example that illustrates the complexity involved in uptake and recontextualization processes can be found in networks formed by a macro genre that embeds micro genres on a single electronic platform. This is the case of the macro genre, a citizen science project, analyzed in one of the case studies.

Finally, the concept of "inter-genre-ality" (Devitt, 2004, 2009) can be applied to the comparative analysis of the generic forms, functions, and social actions enacted by the genres forming a network. Devitt contends that generic forms are "inter-genre-al" because genres "take up forms from the genres with which they inter-act" (Devitt, 2009: 44). For this author, genre form is "part of the fusion of form, substance, and action," and a way of tracing "the visible results *and notable absences of language-use*" (our emphasis added). Devitt further contends that "[a] contextualized treatment of generic form embeds form into its individual, social, and cultural contexts; recognizes generic form as variable individually, synchronically, and diachronically." Moreover, Devitt's concept reinforces the text-first approach that we propose for analyzing genre networks that we explain in Chapter 3. We draw on the assumption that "any complete understanding of genre will need to include the language forms that serve to achieve those purposes and effects, the forms that make generic action happen" (Devitt, 2009: 27). Therefore, we can expect that uncovering fusions of forms will offer a better understanding of processes of knowledge transformation and multisemiotic meaning making and, more broadly, give a better view of the openness of genres and their capacity to interrelate with other genres in new media environments.

In the following sections, we provide a theoretical rationale for the need to move beyond the analysis of individual texts and engage in the analysis of genre interrelations in online environments to understand processes of (re-)entextualization and meaning making in genre networks. We also sketch out the analytical approach to exploring each individual genre within a network, their rhetorical situation, and the attendant social action they aim to accomplish. Once this is done, it becomes feasible to identify and examine the sharedness of substance and form of the genres in the network and by this means the dynamics of intersemiotic relations.

Types of genre collectivity

Recently, there has been growing interest in the exploration of science communication online from the perspectives of genre, discourse, rhetoric, and multimodality. Both genre and rhetorical genre studies have drawn on these perspectives to describe remediated genres and those emerging online for expert and public communication of science. Some examples are online peer reviews, audio slides, video abstracts, video methods articles, registered reports, research blogs, crowdfunding projects, citizen science projects, and microblogging, to mention some of them (Harmon, 2019, Kelly and Miller, 2016; Luzón and Pérez-Llantada, 2022; Mehlenbacher, 2017). The underlying rationale explaining the emergence of these genres is the need to make scientific research open, transparent, and closer to non-expert publics, as also explained in Chapter 1. Taking the genre analytical lens, scholarly research has also been very prolific in describing the rhetorical and language features that characterize digital genres such as three-minute dissertations, TED talks, or science podcasts (Carter-Thomas and Rowley-Jolivet, 2017; Scotto di Carlo, 2014). However, not so many studies to date examine how genres create different "types of collectivity" (Swales, 2004: 5) in the digital environment. We label these various types of collectivity "genre networks," this being a broad umbrella term that embraces other relevant conceptualizations proposed in the literature, such as genre sets, chains, systems, or repertoires (Bray, 2019; Luzón, 2017). Spinuzzi (2003, 2004) and Spinuzzi and Zachry (2000) opt for the term "assemblage" to encompass several categories of genre collectivity—genre sets, genre systems, genre repertoires, and genre ecologies. We have also conceptualized genre ecologies elsewhere (Pérez-Llantada, 2019, 2021b). To avoid overlapping conceptualizations and given that our main focus is exploring analytical procedures rather than developing new taxonomies of types of genre collectivity, in this book we will simply use the term "genre network."

There are further reasons why we propose genre networks. First, because seminal rhetorical genre studies actually refer to the importance of types of "interrelated genres that interact with each other in specific settings" (Bazerman, 1994: 97; see also Bazerman, 2004). Secondly, in studies of professional communication, networks are also described in relation to generic interrelatedness. For example, Yates and Orlikowski (2002: 15) explain that networks are "enacted in some typical sequence" and "composed of a coordinated, interconnected set of communicative actions that together accomplish an interaction." And third, it is our contention that "the increasing generification of administrative and academic life" (Swales, 2004: 4), more marked in today's increasingly interconnected world, requires paying much closer attention to the operation of emerging types of genre collectivity.

Examining intersemiotic relations in genre networks poses several theoretical and methodological questions for genre and discourse analysts. For example, what degree of conceptual depth and what semantic and syntactic features characterize the texts forming these networks to target the imagined audience(s) effectively? And what linguistic, intertextual, rhetorical, and multimodal features make up the form of the genres? And what multisemiotic meanings are created within networks of connected genres and what intersemiotic relations exist between them? While the current analytical approaches to describing single genres in new media environments are very informative, a comprehensive view of interrelations between and among genres, modes and media is needed "to address the *unique combinations* of affordances and constraints introduced by digital media" (Jones et al., 2015: 9, our emphasis).

In discussing digital literacy learning and digitally mediated communication, Jones et al. (2015: 9) adopt Nissenbaum's (2009) theory of "contextual integrity" used in the field of computer science to claim that "texts (and the meanings, social relationships, and identities associated with them) change as they travel from context to context, moving across virtual and physical spaces." The theory of contextual integrity explains that the way texts and utterances are crafted in these processes "depends on the expectations a rhetor has regarding the context in which the text will be interpreted and, more importantly, the imagined audience that interprets the text." But as Jones et al. rightly acknowledge, "the complex overlapping and interested networks of contexts that digital technologies have created makes it much more difficult to maintain contextual integrity." The same applies to the case study research that we present in this book. On the Internet it is actually not always possible to establish clear-cut boundaries to distinguish contexts. We mentioned the concept of the collapsing of contexts in the previous chapter. This means that even if every text is built on contextualized expectations and therefore exhibits contextual integrity, the effects of the collapsing of contexts on the Internet make it difficult to track the exact way in which the wide range of Internet audiences are going to read and interpret them. Although the genres forming a network exhibit their own generic integrity, the different types of interrelations established between or among these genres do reveal a complex overlapping of generic substance and/or form that in turn is manifested at the level of the text in different ways, as we will see in the case study research. In the analysis of the case studies (Chapter 4), we will find several instances of how in Internet environments genres for science communication may hybridize, mix, and blend with other genres, in part as a result of the collapsing of contexts. We also aim to show how these discoursal phenomena are an inherent feature of some of the genres constituting a given network that we analyze. In the scholarly literature we find insightful descriptions of these discoursal phenomena. For example, in science popularization genres, where discourse is recontextualized from the scientific to the journalistic spheres (Motta-Roth and Scherer, 2016: 171) and in para-scientific genres such as citizen science projects and crowdfunding projects. In these genres the discourse merges both formal features typical of scientific discourse and features of science popularizations such as colloquialization features that indicate the influence of speech on written language in online environments (Mehlenbacher, 2017; Pérez-Llantada, 2021a). This is also the case of other emerging genres of science communication in digital environments (see, e.g. Rowley-Jolivet and Carter-Thomas, 2019).

A further complication in determining the generic integrity of connected genres in digital environments is the fact that the form of certain genres in a given network may not yet be fully stabilized. Emerging genres that are relatively new and do not yet show stabilization of generic forms are data articles (Shaklee, 2014), and video methods articles (Hafner, 2018), that appear on the journal platform along with their associated written articles. Both genre types might be considered evolutionary developments of the research article genre, but both exhibit their own generic integrity. It therefore seems apposite to support Bhatia's (2004: 113) contention that the integrity of any generic form should necessarily be viewed as dynamic and flexible, a contention which also aligns with our previous claim of the openness of genres.

Theoretical bases

As in every research study, it is important to establish a theoretical basis for setting the aims of the applied (case study) research. In the present case, we need to define how exactly we aim to address the intersemiotic relations established among and/or within connected genres in online environments. But first, it is crucial to understand multisemiotic meaning-making processes in these systems, what actors are involved in them and what functions language and other semiotic resources perform. Secondly, it is important to trace intertextual links between and among genres. As Bazerman (1994: 97–99) remarks, genre networks "instantiate the participation of all parties [...] embody[ing] the full history of speech events as intertextual occurrences." In our view, the task of the analyst should then be to understand how a given text, which is socio-historically situated, is transformed and re-entextualized in another generic form or forms and/or modes. Thirdly, it is equally important to deepen the multisemiotic meaning making resources integrated into the digital medium, such as the concurrent use of different modes of expression. Along with multimodality, the medium also provides technological resources and tools such as hypertextuality and interactivity features that account for the interrelations between or among the genres of the network. Prior (2009) proposes the notion of mediated multimodal genre systems and analyses multimodality in the "chains of discourse that make up a genre system." Drawing on Lemke's (1998) work, Prior (2009: 17–18) further recognizes the interrelatedness of genres in digital media environments. As he puts it, "literate activity involves multimodal chains of genres." We can establish a possible analogy between Prior's concept of mediated multimodal genre systems and our proposed concept of multisemiotic genre networks which we aim to illustrate with a selection of case studies. Like mediated multimodal genre systems, our conceptualization of genre network should be built upon critical inquiry into those aspects of language use shaped and constrained by the specific (virtual) site or sites of social interaction where the genres of the network are used. In other words, understanding processes of discoursal uptake and recontextualization requires examination of the "trajectories of mediated activity that is, not only in the whole ensemble of discourse production, representation, distribution, and reception, but also in the activity and socialization that flow along with and form that ensemble" (Prior, 2009: 29).

The methodological design that we set forth for the case study research combines a text-first approach with a context-first approach—i.e. a bottom-up/top-down approach or the other way round (Paltridge, 2012; Swales, 2004). We argue that this two-fold approach allows the analyst to describe and explain the results of the analysis and interpret them, taking into account the textual features and the socio-historical situatedness of all the genres forming the network. As we will see in Chapter 4, applying the methodological framework to the selected case studies will entail systematically identifying the linguistic, discoursal, rhetorical, intertextual, multimodal, and hypertextual features of connected genres. It will also involve comparisons of genre, discourse style, and register features across the genres in order to determine whether boundaries between connected genres can be initially set. In addition, it will require analysis of both manifest and generic intertextuality, assisted by corpus linguistics tools, and informed by statistical measures. This analysis will serve to determine the extent to which textual boundaries are blurred and the degree of shared contents and intersubjective positioning across the texts. In other words, it aims to reveal the extent to which the substance of a generic form overlaps with that of another generic form or other forms so as to understand the functionality (or functionalities) and rhetorical effects of such overlapping, always putting the emphasis on comparisons across semiotic modes and media. The outcome of the analysis will be helpful to comprehend better how both the form and substance of interrelated genres reflect new rhetorical exigences that are socio-historically situated.

The theoretical basis of the case study research thus invites us to set out a methodological approach that seeks to move from the description of single genres to the enquiry into intertextual and hypermodal features with a view to better grasping genre connections and intersemiotic relations within genre networks. Bawarshi (2010: 199–200) remarks that genre uptakes "are tacit, deeply ingrained, and ideologically consequential" because the rhetor needs to make decisions as regards "what to take up, how, and when." The analyst should then be able to examine discoursal uptake to assess critically how genres are repurposed and refocused in another genre, even at times in more than one genre, and interpret genre connections taking into account the situational and socio-historical context of such genres. Two influential and closely related theoretical concepts are thus worth considering for examining intersemiotic relations in digital genre assemblages: "uptake" and "intermediary genres." Freadman (2002) borrows the notion of uptake from speech act theory to put the emphasis on the connections between texts. From an analytical perspective, this implies moving from the merely textual to the non-textual part of the utterance, or the social and cultural contexts of genres (Paltridge, 2012: 78; see also Swales, 1990, 2004). In discussing Freadman's conceptualization of uptake, Artemeva (2004: 4) further stresses the relevance of the non-textual part of the utterance and also advocates the need to examine "how multiple genres relate and interact through uptake" to better grasp rhetorical agency and, from the perspective of writing pedagogy, "the rhetor's ability to reshape and manipulate genres to suit certain rhetorical situations" (Adam and Artemeva, 2002: 181). It follows that in investigating uptake, the analyst should identify the identity of the rhetor within a socio-historical situation, the motives that he/she has for creating a new text, the strategies that he/she uses to transform the previous text, and re-entextualize it in a new text, as well as the perspective and viewpoint that he/she takes towards the new text's contents. At times, we may find genre networks where the same rhetor or rhetors create all the genres of the network, as case studies 3 and 4 will illustrate. At other times the networks contain genres whose discourse is (re-)entextualized by different rhetors in other genres, as in case studies 1, 2, and 5.

To understand discoursal uptake and processes of recontextualization of scientific knowledge across different semiotic modes, the analyst needs to know the ways in which the contents and views expressed in one genre are reinterpreted, compressed, or expanded in the new text. In other words, the outcome of the rhetor's or rhetors' decision(s) as to how to transform the contents and perspectives expressed in one genre into another genre and to what extent the substance and form of the source text should either remain or change, represents the non-textual part of the utterance. Bawarshi (2010: 199) rightly acknowledges that discursive uptakes are "complex, often habitualized, socio-cognitive pathways that mediate our interactions with others and the world" because they may be tacit and/or ideologically laden. Indeed, tacit meanings and textual silences can also be richly informative when interpreting how the situatedness of a generified social action is entextualized through language, discourse resources, and multisemiotic modes of expression.

In addition, it is important to discern how the rhetor or rhetors construct(s) their imagined audiences—the specific conceptualization that rhetors have of their audiences in a mediated conversation (Marwick and Boyd, 2011)— as the imagined audience determines the choice of language forms and the creation of relations among genres. If we move the focus from knowledge production to knowledge reception—or digital text receptivity—the analyst may find it useful to adopt Bakhtin's (1981) concept of the "threshold of interpretation" to explain how the processes of re-entextualization in subsequent discursive uptakes may set up "thresholds" or specific entry points to access information depending on the rhetor(s)' imagined audience.

Genette (1997) also uses the concept of threshold of interpretation to define para-texts as the range of textual resources that accompany a text, enhance its contents, and give perspectives of the contexts and agents related to such text (Pérez-Llantada, 2021b; see also Chapter 3 for further discussion). Case study 1, for instance, explores some para-texts in an enhanced publication. The terms threshold of interpretation and para-texts are useful analogies for understanding how genres working in conjunction with other genres in web environments do not always act as self-contained artifacts. Even if certain genres in a network may have a high degree of conceptual depth—say, they exhibit both semantic and syntactic complexity—such conceptual depth may not necessarily restrict the access to such text even if they are not the intended audience. In other words, the presumed threshold of conceptual depth (entailing certain knowledge expertise) initially established by each genre may be crossed on the web by different kinds of readers or imagined audience(s). On the web, readers may even move back and forth between genres characterized by different degrees of conceptual depth and access texts specifically targeting expert audiences, non-specialist readers, diversified audiences, and general publics. In short, in the virtual space thresholds are not clearly perceptible and para-texts may mediate the knowledge base required to understand a text.

The genre analytical lens

Genre analysis is a very powerful analytical tool for understanding how genres enact social actions and what discourse and hypermodal resources they use in digital environments (for further discussion see Mehlenbacher and Mehlenbacher, 2020). Basically, applying the genre analytical lens involves both description and explanation of the rhetorical information organization of a genre exemplar—i.e. the genre sociocognitive schemata—and its main language features. This paves the way to explaining why the producer of the text—i.e. the rhetor or rhetors—use(s) language and other semiotic modes in particular ways (Bhatia, 1996: 39). Rhetorical genre analysis enquires first into the rhetorical situation, or the situational context that motivates the production/reception of a genre exemplar, to later explain the action that the genre accomplishes and the social and historical aspects that determine its generic form (see Swales, 1990: 42 for further discussion). Flowerdew (2002: 91–92) further explains that while the rhetorical approach is "situation-first," which looks to the text "to interpret the situational context," the linguistic approach is "text-first," which means that it "looks to the situational context to interpret the linguistic and discourse structures." Swales (2004: 72–73) elaborates on Flowerdew's (2002) distinction between text-first *vs* situation-first and proposes two possible procedures for conducting genre analysis, that he calls "text-driven" and "situation-driven." As this author explains, the starting point of the former procedure is the analysis of the structure, style, content—i.e. the semantic profile of the genre form—and purpose of the exemplar, and this enables the analyst to interpret the context and, if apposite, to "re-purpose" the genre's communicative goals. This is the approach taken in this book. In a situation-driven procedure, the starting point of the analysis is the identification of a communicative situation that leads the analyst to re-purpose the genre's intended communicative goals by taking into consideration the goals, values, and material conditions of the groups involved in the situation, their horizon of expectations and their genre repertoires.

Applying a situation-driven procedure, rhetorical genre analysis first captures the socio-historical context of genres in order to understand, for instance, why digital genres for expert communication of science have evolved—e.g. registered reports, online laboratory notebooks, or graphical abstracts, among others—and why para-scientific genres like blogs and microblogging have emerged in media environments (Harmon, 2019; Kelly and Miller, 2016; Mauranen, 2013; Mehlenbacher, 2019a; Miller and Kelly, 2017). Rhetorical genre studies have mainly shed light on the multiple and intersecting socio-historical factors of the science–policy interface that act as drivers of generic evolution and change. The most influential have been "fast-expanding digital publishing" (Mauranen, 2013) in the scholarly world, and the social and science policy agendas such as those of the OS movement. Rhetorical genre studies have investigated a wide range of emerging genres online. One example is the laboratory notebook, a traditional, occluded genre that has been remediated or moved online to support the sharing of data and research protocols among researchers. As a result of the process of remediation, and as part of its generic innovation, the genre online includes emotive and informal elements in fulfilling its knowledge-sharing action (Carter-Thomas and Rowley-Jolivet, 2017). Another example of how analysis of the socio-historical context can help clarify the nature of emerging genres is the open peer review, one of the genres of the network described in Case study 2. Ross-Hellauer (2017) explains that in the digital environment, this genre has evolved into different types of peer review—e.g. open reports published alongside articles or reviews where the wider community, and not only invited experts, can contribute. This is done with a view to enhancing the rigor of methodological research processes. Mauranen (2013) also notes that the innovations in digital publishing that result from the strong impact of technologies on researchers' activity very likely account for the emergence of blogs, that she describes as an emerging form of communication that allows researchers to share their doubts about their ongoing research using social media.

On the other hand, text-first studies, also called text-linguistic studies, zoom into the recurring linguistic and discoursal resources that characterize emerging genres in digital environments. Most are actually genres that embed audiovisual innovations. Here, it is worth recalling that these studies have analyzed the rhetors' specific choice of language features and discourse styles to understand the ways in which specialist knowledge has been recontextualized in the digital medium. These studies have shown that scientists use language in particular ways to address the polycontextuality of the digital medium and take advantage of the functionalities of its hyper-linking affordances. For instance, Orpin (2019) reports that certain linguistic resources signaling authorial stance, such as first-person pronouns and reader pronouns, are deployed in debates about vaccine safety on Twitter based on expert epidemiology reports. The appropriate choice of language features in processes of genre remediation and knowledge recontextualization is fundamental to achieve the genre's rhetorical goals. One possible effect could be the use of self- and others' references—first- and second-person pronouns—in genres such as laboratory notebooks and in spoken genres such as three-minute thesis dissertations, author videos, and podcasts (Rowley-Jolivet and Carter-Thomas, 2019). The scholarly literature has also reported aspects of generic hybridization. Some examples are the presence of grammar and discourse style features associated with informal registers characteristic of the language of conversation in genres such as crowdfunding science projects (Pérez-Llantada, 2021a), the use of modal verbs in professional (remediated) genres online, such as open peer reviews (Breeze, 2019) or the use of personal pronouns along with attitude markers to establish proximity with the audience and create a professional identity in genres such as blogs (Hyland, 2018) and audiovisual/audio slide presentations (Yang, 2017).

Although, as Swales (1990) explains, both text-first and situation-first procedures can be combined, we prefer to apply the former approach in all the selected case studies, for the sake of consistency. For this reason, we will first look at the recurring features of texts in different textual layers, the linguistic, discoursal, rhetorical, multimodal, and hypertextual. Secondly, we will trace the semiotic interrelations established among connected texts, taking into account the situational context in which the texts are produced, transformed, and taken up by one or more rhetorical actors. This two-fold approach aligns with the interpretations of rhetorical situations provided by studies that have analyzed genre uptake in different types of collectivity, such as genre sets, genre chains, or genre systems (Bray, 2019; Luzón, 2017; Reid and Anson, 2019; Smart and Falconer, 2019). These studies emphasize the value of the "web-based genre assemblage" construct to assess critically how scientific contents are created, transformed, and reinterpreted. Some of the specific questions that we investigate applying the genre analytical lens are the following: what degree of semantic complexity (and conceptual depth) characterizes the different texts forming a given network? How is multisemiotic meaning making assembled within the network and for what purposes? Or what are the functionalities of the multimodal and hypertextual affordances in the network? These are the kinds of questions that we aim to answer by applying the genre analytical lens to our case study research, to which we turn below.

Case study research

We have selected five case studies to illustrate genre connections online and processes of discursive uptake and recontextualization. In these studies, we analyze, among others, three genre types according to the kind of discourse that they represent. The first type is representative of scientific discourse, or what Kelly and Miller (2016) conceptualize as internal genres of science communication, such as journal articles and abstracts, video methods articles, and other expert genres such as journal editorials and editors' decision letters. The second type of discourse is that represented by para-scientific genres that, as stated previously, combine research-based discourse and informal discourse to target diversified audiences, such as blogs, tweets, and citizen science projects. These genres are therefore exemplars of science popularization discourse. The third type includes genres that belong to other orders of discourse, such as institutional discourse found, for example, in institutional websites.

We have selected genres representing scientists' professional communication practices, ranging from enhanced publications resulting from processes of genre remediation such as articles and abstracts online and open peer reviews, as well as genres accompanying these enhanced publications such as short summaries, impact statements, and lay or plain language summaries. The case studies also include exemplars of other multimodal generic innovations such as video methods articles, interactive tutorials, tasks and field guides, video-based interviews embedded in web portals and blogs, as well as tweets in social media. With this palette of genres, we aim to explain how these different genre types interact and respond to the rhetorical exigences posed by their situational and socio-historical contexts. We also aim to illustrate how multimodal, hypertextual, and interactivity affordances enable the rhetors to enact each genre's attendant social action. In exploring the intersections of these genre types, we aim to gain a better understanding of intersemiotic relations of connected genres in digital environments. Table 2.1 summarizes the specific goal of each case study and lists all the genre exemplars included in each network of genres. The specific networks analyzed in each case study exploit the digital affordances in a different way and hence the focus of analysis will vary across the cases. Although it should be acknowledged that it may not always be possible to generalize from the findings of these illustrative case studies, the value of these exploratory cases lies in the fact that they can capture important aspects of the complex multisemiotic meaning-making processes involved in connected genres online.

In Chapter 1 we briefly outlined the case studies in terms of contents or thematic orientation to show how they all address scientific contents related

Table 2.1 Specific goals of case study research

Case study #	Genre exemplars included in each network	Specific goals
Case study 1. High-stakes science	Web page text (including its embedded video); journal editor's short text and editorial; article and abstract; science blog post	To describe genre connections in a sequential discoursal uptake of institutional, scientific and science popularization genres
Case study 2. Science gatekeeping and add-on genres	Article online; decision letter and authors' response to reviewers; impact statement: digest (or plain language summary)	To describe genre connections in a chain of open access and open science genres for disseminating scientific research
Case study 3. Science reproducibility genres	Video methods article; associated written methods article; table of materials	To describe semiotic interrelations in a multisemiotic genre system for reporting scientific methods and methodological procedures
Case study 4. Participatory science genres	Citizen science project homepage; Overview page (including embedded audiovisual interview); About page (Research, Team, Results, FAQs); Classify page (Task, Tutorial), embedded Field guide	To describe semiotic interrelations within a single macro genre that contains modular texts and several embedded micro genres
Case study 5. Para-scientific and social media genres	Article online; online science news article; tweets	To describe discursive uptake and recontextualization of a genre for professional communication of science in social media genres

to two SDGs, as shown in Table 1.1. We would like to highlight here that the choice of these case studies also aimed to illustrate different possibilities for analyzing genres online, as detailed in Table 2.1. More broadly, with this case study research we want to showcase the openness of connected genres online, their functional versatility, as Giltrow and Stein (2009) call it, and their multimodal (verbal and audio/visual), hypertextual (web-based) and interactivity features. Moreover, from the analysis of these case studies we expect to make sense of the socio-historical contexts in which these different genre exemplars are produced and received.

Bhatia (1996: 40) explains that "[g]enre theory tends to give a grounded or what sociologists call a 'thick' description of language use rather than a surface-level description of statistically significant features of language, which has been very typical of much of register analysis." In this book, however, we tend to give equal value to both types of description, stressing how they complement each other. We believe that both mutually inform one another and thus provide a greater understanding of the intersections of genres, semiotic modes, and media in relation to multisemiotic meaning-making processes online. We argue that viewing genre analysis in this way can integrate both a bottom-up enquiry into the regularities of a given text and the use of lexicogrammar, discourse, and register features with a top-down examination of the socio-rhetorical situation as reflected, for example, in horizontal and vertical intertextuality features and in features of interdiscursivity. Aligning with Bhatia (1996: 51), we consider genre analysis as "narrow in focus and broad in vision" and thus an effective analytical approach to address generic versatility in digital environments.

Table 2.2 summarizes the main text-internal properties that we trace in the different genre exemplars forming each network—lexical, grammatical, discoursal, multimodal, hypertextual, and rhetorical. It also lists the text-external properties, or sociocognitive and cultural factors determining or constraining the use of certain text-internal properties in each genre. These synergistic layers of textual analysis also take on board Bateman's (2014: 251) assertion that "[d]ifferent semiotic modes generally employ different kinds of discourse semantics." Like Swales (1990: 6), we recognize that textual knowledge remains insufficient. Hence, with this approach, we expect to cover "how texts organize themselves informationally, rhetorically and stylistically" and how each discursive uptake brings with it specific contextual (text-external) variables. Applying this analytical approach to each individual genre exemplar of the network and comparing the resulting analyses will, in our view, give us "a more dynamic explanation of the way expert users of language exploit and manipulate generic conventions to achieve a variety of complex goals in response to recurring and changing rhetorical contexts" (Bhatia, 1996: 55).

As shown in Table 2.2, the text-first approach conflates several perspectives, namely the linguistic, the discourse, the multisemiotic, and the socio-rhetorical. Such a multiperspective analytical approach allows us to enquire into the formal features of genres and to compare them with those of other genres connected with them in various ways. While this approach mainly draws on (critical) genre and discourse analytical methods, it also nurtures from other analytical procedures such as those used for analyzing register features. We understand that this helps to grasp "the *specialized functions of the texts* and their surrounding social contexts" (Flowerdew, 2001: 23, our emphasis). We also advocate the integration of several analytical procedures by drawing on Devitt's (2009: 27) contention that "any complete understanding of genre will need to include the language forms that serve to achieve those purposes and effects, the forms that make generic action happen." It is our initial assumption that identifying the forms that make connected genres

Table 2.2 Proposed text-first approach (adapted from Bhatia, 2017b)

Type of features	Textual layer of analysis
Text-internal features	Rhetorical (how texts are organized rhetorically) • overall textual macrostructure • move structure • rhetoric and argumentation Linguistic (how texts are organized informationally) • lexical density and complexity • syntactic complexity • phraseological profile Discourse style and register (how texts are constructed stylistically) • register features • agency and self-representation • discourse strategies • stance and evaluation Multimodality • types of visuals • functions of visuals • embedded visual elements • visual-verbal interrelations Intertextuality and interdiscursivity • manifest intertextuality • genre mixing or blending • interdiscursive colonization • resemiotization Hypertextual navigation options (functions of links)
Contextual (text-external) features	• situational context • socio-historical (rhetorical) context • social actors (or rhetors) • participant relations • participants' contribution to the process of genre construction • social structures (professional/institutional/disciplinary practices and constraints) • underlying ideologies, beliefs, and values

accomplish their actions leads to fusions, or intersections, of language and generic forms that reveal complex meaning-making processes in genre networks online. Like Freadman (1994: 45), we view digital science communication as "the product of the interaction of a variety of 'languages' or semiotic systems, not necessarily homologous with any other."

In this chapter we have conceptualized both genre and connected genres online which entail multisemiotic meaning making in the form of genre networks. We have also given our main theoretical and analytical rationales for analyzing and critically interpreting networks of genres that communicate science in digital environments. We have explained that we mainly draw upon (critical) genre analysis together with discourse and rhetorical analysis to address the methodological challenges posed by the openness and interconnectedness of genres in digital environments. We have also briefly presented the case study research reported in Chapter 4, that aims to illustrate several intersemiotic relations among genres. In the following chapter we elaborate further on the specific analytical procedures that we employ to describe the functional goals of connected genres supported by Web 2.0.

3 Layers of textual analysis

Connected genres online forming different types of genre networks, the main object of enquiry in the case study research of Chapter 4, requires investigation into the way language and other semiotic forms of meaning making are deployed in processes of genre uptake and knowledge recontextualization and resemiotization in web environments. In order to identify the semiotic interrelations established between or among the genres of a given network, like Bhatia (2004) we propose the integration of several analytical procedures, namely a multianalytical approach. We complement (critical) genre analysis (Bhatia, 2004, 2017a; Swales, 1990, 2004), with analytical procedures used in rhetorical genre studies (Bazerman, 1994, 2004a, b; Bhatia, 1993) and register analysis (Biber and Conrad, 2020), and studies of multimodality (Bezemer and Kress, 2008; Jones and Hafner, 2012; O'Halloran et al., 2019; Van Leeuwen, 2005). In complementing the linguistic and rhetorical analyses of genre networks with i) quantitative, corpus linguistics tools queries, ii) close reading of the texts and their semiotic elements, and iii) manual coding procedures for qualitative analysis, we aim to interpret semiotic interrelations based on descriptive statistics and grounded values. We would argue here that this point of departure supports the analyst when retrieving, classifying, and interpreting textual data expressed in different semiotic modes and genres. In this chapter we summarize those simple analytical procedures that can empirically inform and guide the textual analysis of the genres involved in the network and, more importantly, of the intersemiotic relations established between or among them. Below, we explain some possible ways of applying these procedures to gain richer insights into the lexico-syntactic, discoursal, intertextual, rhetorical, multimodal, and hypertextual profiles of the genre networks investigated.

The main reason behind our integral analytical approach is to add rigor to the analysis and interpretation of the complex multisemiotic meaning making in connected genres in a digital environment. The approach intends to capture the openness and functional versatility of genres in these environments and offer empirically based commentary on concepts regarding science communication online, for example the blurring of

DOI: 10.4324/9781003147732-3

thresholds of interpretation, the purported sequentiality of discursive uptake, and the appropriation of discursive features of one genre by another genre (or interdiscursivity). These concepts have not been dealt with in the examination of genre interrelations in digital science communication and are thus worth addressing further to understand the effects of the collapse of context (Brandtzaeg and Lüders, 2018) in Web 2.0. The multianalytical approach also aims to validate empirically the view that each genre should not be conceived of as a self-contained construct containing one or more semiotic modes but rather as a construct that interrelates in different ways and by different means and modes with other genres online. Evidence of different types of intersemiosis could demonstrate the openness of genres and yield more robust interpretations of the ways in which processes of genre uptake and recontextualization eventually create a rich polyphony of voices and perspectives towards issues of science in web environments.

Through case study research, we try to show why the analyst needs to critically assess the implications of "how the various discourses interact, what modes are involved, and what is communicated," in the words of Myers (2003: 267). As we will see, discourses on the web complement one another and interact in various ways. The web creates and sustains different rhetors' identities, views, and perspectives, which supports Bhatia's (2004: 155) contention that "the focus of textual analysis needs to shift to 'more complex and dynamic aspects of discourse construction and interpretation'." This is why we propose to investigate discourse construction by describing the lexico-syntactic, discoursal, intertextual, multimodal, and hypertextual layers of connected genres in a comparative manner. This integrated analytical approach is summarized in Figure 3.1.

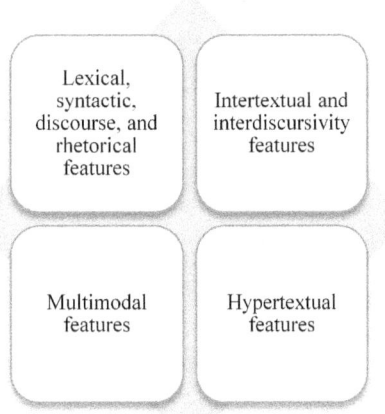

Figure 3.1 Layers of textual analysis in genre networks online.

All the case studies explore these different layers of analysis. However, for the sake of illustrating the multifaceted nature of intersemiotic relations in connected genres, each case specifically places a greater focus on the analysis of one (or more) of these layers. We show how such analysis can be informed and/or enriched by data that are statistically validated or that are grounded and by data that correspond to other layers of textual analysis. In the following sections we summarize the different analytical procedures that we apply in the case study research. In stressing the advantages of analytical complementarity, we align with Flowerdew's (2012: 175) contention that corpus and discourse approaches, with "their inherent differences in epistemologies and methodologies" can establish fruitful synergies while solving the limitations of each approach.

Genre analysis and corpus tools

Genre analysis is a very strong analytical approach to studying how genre exemplars are organized rhetorically and constructed stylistically. Close analysis of the overall information organization macrostructure (the macro level), the rhetorical move/step organization (the meso level, as defined by Swales, 1990), and the language features (the micro level of the text) makes it possible to grasp the construction of knowledge in processes of knowledge transformation, recontextualization, and resemiotization. Enhancing the validity of the genre analytical approach, corpus linguistics tools can assist the analyst in the identification of the main linguistic elements of each text and its immediate co-text (Stubbs, 2001) and the comparison of these elements across the texts in the network. Descriptive statistics measures such as type/token ratio (TTR) and standardized type/token ratio (STTR), measures of lexical density and lexical and syntactic complexity, and other basic descriptive statistics such as frequency of words and keywords or word grams, clusters, and concordances, provide the analyst with helpful information about the salient linguistic, discourse, and register features of the texts included in the network. Corpus searches also prove revealing when examining stretches of discourse that contain meanings that may not be explicitly stated in the text (Stubbs, 2001), a similar procedure to that used in corpus-based discourse studies (see, e.g. Flowerdew, 2001, 2012). The corpus-informed approach that we propose, based on automated searches of words and word combinations, thus helps the analyst in making decisions as to how to proceed with the move/step analysis or, as another example, how to interpret traces of intertextuality and features of interdiscursivity across the genres and modes involved in the network. We also view it as a more rigorous basis for the comparison of the structures and functions of clause- and phrase-level features across genre exemplars (Loi and Evans, 2010; Pérez-Llantada, 2015). As also shown in the case study research, corpus-informed genre analysis also allows the analyst to identify features of self-/others' representation(s) as well as the underlying ideologies that the text creators reflect in their discourses by tracing the presence of the relevant linguistic resources. In sum, corpus linguistics methods used in genre analysis uncover language features that point to the situational context in which a genre exemplar is produced, used, and received.

Existing commercial software such as Wordsmith Tools (Scott, 2008) and free software such as Kfngram (Fletcher, 2002–2007), Antconc (Anthony, 2019), FireAnt (Anthony and Hardaker, 2017), and #LancsBox v6 (Brezina et al., 2015, 2020), among many others, can be very helpful for doing small-scale corpus searches and getting results. For discourse analysis purposes, existing qualitative software such as Atlas.ti and NVivo can also be used to search for words and collocates, to compute descriptive statistics and, above all, to code linguistic elements, rhetorical moves, multimodal elements, and other semiotic resources that are worth examining. The outcome of these procedures becomes a good starting point from which to move on to the examination of types of interrelations, such as code-document co-occurrence (that enables the tracing of relations between code grounded value and number of quotes in a given text), and code co-occurrence (or occurrence of two codes in the same textual fragment, which allows the identification of concomitant use of salient features). In this book we explore language variation in processes of recontextualization and resemiotization in small datasets of connected texts ranging from 5,000 to 10,000 words approximately, depending on each genre network. In each case study we treat these small datasets as corpora (and subcorpora at times), understanding "corpus" as a collection of texts that can be used to study aspects of language. The size of these corpora is relative to the frequency of the textual features that we track and/or examine and proves to be sufficient to establish cut-off points and frequency levels for comparing meaning making in genres and modes, finding out the distinct details of each text worth commenting on, and assessing the extent to which each genre exemplar shares form and substance with the remaining genre exemplars in the network. The size of the selected networks is also appropriate for manual coding and subsequent categorization of linguistic and multimodal resources using software for qualitative analysis.

Measures based on descriptive statistics and groundedness are also very informative when comparing the nature of language in interrelated genres that represent different types or modes of discourses (i.e. descriptive, expository, narrative, and argumentative) and describing different degrees of conceptual depth in processes of "transmedial gradation" (Maier and Engberg, 2019). Both aspects are crucial for understanding knowledge recontextualization and resemiotization in digital media environments. These measures also indicate the range of vocabulary used to describe notions and concepts of science and express opinions, values, and viewpoints about them. In addition, they can be employed to understand why texts in the genre network may show greater or lesser lexical variety, for example written *vs* spoken texts or verbal *vs* oral modes. The view of grammar as a language variable determining the discourse style that characterizes a given text (Biber and Gray, 2010, 2016; Biber et al., 1999) and, specifically, the concepts of complexity, elaboration, and explicitness, and descriptions of phrasal *vs* clausal discourse styles to trace register variation, are also of considerable heuristic value when interpreting intersemiotic relations within genre networks. These language aspects enable the analyst to look at language features within clause boundaries (i.e. what Miller (1984) refers to as form or syntax) to identify the rhetorical exigences that influence or determine the semantics of the texts (what Miller conceptualizes as the substance of genres) in these networks. Alongside aspects of grammar, lexical density is a particularly useful measure to compare the degree of conceptual depth entailed in knowledge entextualized in an expert-to-expert text and in its subsequent genre uptake intended to target multiple publics. Moreover, these measures assist the analyst in corroborating the linguistic features specifically associated with different register types such as those typical of journalism, academic prose, and conversation (Biber and Conrad, 2020; Biber and Gray, 2016; Biber et al., 1999) and in detecting language variation in, say, the use of style features and aesthetic preferences.

In many ways, the integral analytical methodology is a rather similar approach to that taken in corpus-assisted discourse studies, where corpus-retrieved data are used to establish the lexical and syntactic profiles of the texts by enabling systematic procedures in data handling and data analysis. Drawing on Sinclair's (2008) theory of language as phraseology, Römer (2010: 96–97) remarks that focusing on phrases rather than on isolated words "preserves the integrity of the text" and thus represents a sound approach to describe meaning in texts insofar as it "provid[es] insights into meaning creation in the discourse." As Römer (2010: 214) further notes, the extraction of contiguous word combinations, called wordgrams, collocations and clusters in combination with concordance analysis can be particularly useful to identify the pragmatics of modality and aspects of stance and evaluation in texts (Hunston and Francis, 1996; Hunston and Thompson, 2000; Römer, 2008). The analysis of discourse pragmatics features across connected genres and modes reveals the different ways the rhetors position themselves towards the utterances expressed in their texts. As Hyland (2010: 116) puts it, writers "display both authority as an expert and a personal position towards issues in an unfolding text." As we will see in the case study research, tracing linguistic resources of authorial stance and engagement will uncover the way the rhetors construct arguments to convey their views and perspectives. Identifying these resources paves the way to understanding the what, how, and why of processes of discoursal uptake and recontextualization in networks of genres. In short, these analytical procedures can successfully inform the analyst about the overall pragmatics of the discourse, already explored in other digital genres (Büchi, 2016; Caliendo, 2012; Scotto di Carlo, 2014), and guide him/her in the subsequent (critical) genre analysis of the texts.

Genre analysis and intertextuality

Broadly speaking, the concept of intertextuality refers to the relations established between texts that may take the form of adaptation of previous works or citations of previous works and/or attribution to previous works, thereby assuming that every text is "multivoiced" (Kristeva, 1980: 37). In this section we provide a brief summary of how Bakhtin's (1981, 1986) influential theory of intertextuality in discourse and speech acts can be applied to the analysis of intersemiotic relations in genre networks. As we have argued elsewhere, Bakhtin's theory of intertextuality has been applied extensively not only in the field of literary studies and criticism, but also in other non-literary fields of study, such as discourse analysis, critical discourse analysis, and (critical) genre analysis (Bazerman, 2004a; Bhatia, 2004; Pérez-Llantada, 2021b). In (critical) discourse studies, a distinction is made between manifest intertextuality and tacit intertextuality. While the former is defined as the explicit reference to previous texts in a given text, the latter is used to refer to non-explicit references from which assumptions can be made (Fairclough, 1992: 104). Both types of intertextuality have been investigated extensively in discourse studies to assess the influence of previous texts (or pre-texts) in discourse and understand explicit assumptions and tracing tacit or implicit ones.

In the case study research, we follow Fairclough (2003: 192) and Bazerman's (2004a, b) approaches to intertextuality. We investigate first the relevant voices (or other texts), both those included or "significantly excluded." We then explore the rhetorical section in which these relevant voices or other texts are placed and also the form they take, for example quotation or citation. In both cases, the previous texts or pre-texts are explicitly present in the text. We also investigate how the traces of others' texts (and voices) are re-entextualized linguistically, rhetorically, and stylistically and, given that our focus is on interconnected genres, how and why the contents and the rhetors' views are transformed in different ways, for example by reaffirming the claims made in the previous text, or countering the arguments made in it. We also analyze how and why the rhetors' voices and perspectives are refocused using different semiotic resources. Enquiry into these different aspects of intertextuality can give us a view of text-internal intertextual relations. In the case study research, we will employ frequency lists, pairwise comparison of keywords, and Key Word In Context (KWIC) analysis as a point of departure to trace content (i.e. manifest) intertextuality and interpret the results from the genre lens, above all, along the lines proposed by Bazerman (2004b). Specifically, to interpret instances of manifest intertextuality, we draw on recurring collocates and concordance lines. By collocates we understand the nearby words accompanying a key word that convey salient meanings (for example, looking at types of collocates such as lemmas, part of speech and synonyms) (for further discussion see, e.g. McEnery and Hardie, 2011).

We complement the corpus query searches with close reading and the use of tools for qualitative analysis. These procedures are applied systematically to trace intertextual links across the genre types comprising the generic network. When the texts forming the assemblage are sequential (i.e. when they follow a chronological order), it is easy to identify what text has been taken up in a subsequent text or post-texts. However, as we will explain later, not all the networks illustrate sequential chains of discoursal uptake because, as Spinuzzi (2004) notes, genres interconnect with other genres in complex ways. The reader will find an example in Case study 4. In the case studies we necessarily take one genre as a starting point to describe discourse uptake and the process of knowledge transformation, but we acknowledge that this process is not always sequential in some cases. A first-hand example is the research article in an enhanced publication whose contents can be recontextualized in more than one genre, for example, an impact statement, an audio slide presentation, a podcast interview, a news report, or a tweet, to name a few.

Another very influential literary approach to intertextual analysis that we draw upon to understand multisemiotic meaning making and intersemiotic relations in connected genres is that proposed by Genette (1997). Specifically, we borrow Genette's concept of para-text as being any text (either pre- or post-text) that is associated with a single text and that contributes a polyphony of voices that may influence the interpretation of the source text. An example of a para-text in a digital environment would be the editors' short texts introducing online articles in electronic journals, as shown in Case study 1. In addition, Genette explains that a literary work is accompanied by other surrounding texts that he conceptualizes as peri-texts and that bring in multiple voices or viewpoints. Each of these texts acts as a threshold "that offers to anyone and everyone the possibility either of entering or of turning back." Genette refers to this threshold as "a zone not just of transition, but of *transaction*" (Genette, 1997: 261; our emphasis).

Briggs and Bauman (1992: 132) explain that intertextuality is a prototypical feature of genres related to scholarly writing that determines the "textual open-endedness" of such genres. In this book, we conceive of intertextual analysis as a way of enhancing our understanding of the social and ideological factors that underlie the semiotics of genres working in conjunction with other genres in a digital environment. As we will see in Chapter 4, in tracing features of intertextuality it is possible to address critically the motivations for the presence of previous texts and their significance for the construction of the subsequent text. This analytical procedure allows a better understanding of how and why the rhetors' voices and perspectives towards issues of science vary and why the underlying ideologies may differ across the texts. Investigating intertextual relations— also signaled by hypertextual links—allows the analyst to track "strategies for minimizing gaps between texts and generic precedents with strategies for maximizing such gaps" (Briggs and Bauman, 1992: 131). From a pedagogical standpoint, understanding intertextual relations within genre networks online can inform LSP and EAP teaching practice, particularly formal instruction in composing digital genres. Bawarshi and Reiff (2010: 50–51) underline the value of addressing aspects of intertextuality in ESP genre pedagogy by arguing that knowledge and understanding of intertextuality help the target learning group become conscious of the functional versatility of genres.

Because in this book we aim to describe networks of genres rather than single genres, we deem it necessary to analyze both text-internal intertextuality (i.e. intertextual material cited or referred to in a given text) and text-external intertextuality (i.e. shared-ness of generic substance or form between two or more genres of the network). We also add a further facet of text-internal intertextuality to address the case of micro genres embedded in a macro genre. In all these cases, we draw on keyword (KWIC) analysis as a starting point to identify and analyze manifest intertextuality and establish the conceptual overlapping at the level of semantics between or among the genres forming the network, or between or among the micro genres embedded in a macro genre. In addition, we use intertextual analysis to trace the processes of recontextualization and multisemiotic meaning making. This analytical approach has already been used for the analysis of intertextual chains in genres such as textbooks (Lähdesmäki, 2009) and applied in mediated discourse analysis of intertextual relations between discourses in different semiotic modes and media (Scollon, 2008).

We take a similar analytical approach to the study of interdiscursivity, conceptualized as a genre's appropriation of discursive resources from other genres (for further discussion, see Bhatia, 2004). In fact, Lee (2008: 91) points out that intertextuality and interdiscursivity should not be addressed separately, as they are "part of the same phenomenon: that language is never created fresh and from scratch, but borrows, repeats, quotes, implies and alludes to prior texts and prior ideas." Adhering to Lee's point, along with intertextual features we aim to identify features of interdiscursivity and trace the appropriation of discourses (e.g. promotional, educational, etc.) in texts representing other types of discourse. Both intertextuality and interdiscursivity have been considered key features in popularization genres (Motta-Roth and Scherer, 2016) and para-scientific genres (see, e.g. Mauranen, 2013 for the case of research blogs). Enquiry into both features will support our interpretation of genres in relation to the historical, situational, and contextual factors of the texts (Paltridge, 2012). We believe that examining the construction of value and the rhetors' positioning towards it and the different rhetors' voices—i.e. the way they construct value and the way they (re-)entextualize their viewpoints and project an ideology— adds rigor to explanations of how the situational and socio-historical contexts of the genres constrain the processes of re-entextualization and resemiotization.

Genre analysis and multimodality

Computer-mediated communication is fundamentally multimodal (Herring, 2019). In the digital environment, text producers can combine a variety of modes (verbal, aural, visual) to convey meaning, and integrate several elements (e.g. text, images, video, and audio files) in a single platform to create "hybrid" multimodal genres (Adami, 2014; Hafner, 2018). The affordances of the digital environment facilitate the complex interweaving of various modes, the connection between different media on a single or several platforms, and the reuse of elements in different digital documents. Any media type can be linked to another or embedded in a different platform (Adami, 2014), all of which involve processes of recontextualization and repurposing. Therefore, multimodal analysis is crucial for understanding the linguistic and contextual features of emerging audiovisual genres online such as video method articles, of hybrid genres such as research blogs or crowdfunding proposals that embed or connect to other genres and media through hyperlinking, and of genre networks that involve various media and modes.

In the case studies presented in this book we examine how various modes and genres are combined to achieve particular purposes in digital science communication: (i) audiovisual genres, such as video method articles, which are embedded in the hypermodal platform of the online video journal; (ii) enhanced publications, where linguistic resources may be co-deployed with tables, graphs, charts, images, or embedded videos (e.g. video abstracts); (iii) genres for public understanding of science and public communication of science, which may embed photographs or videos (e.g. author pitches, animated videos); and (iv) social media genres such as microblogs, where language is usually co-deployed with images, animations, embedded videos, or emojis.

Our approach to multimodality and multimodal analysis draws upon seminal multimodal studies (Kress, 2010; O'Halloran, 2009) and studies of genre and discourse analysis in LSP (Jones and Hafner, 2012; Paltridge, 2012). From the perspective of Multimodal Discourse Analysis (MDA), in a certain context a variety of modes, both linguistic and non-linguistic (e.g. writing, speech, still image, moving image, color, sound), are integrated to create meaning. Modes have different affordances (distinct potentialities and limitations) and are therefore used to convey different meanings, but no one mode has more potential than the others (Jewitt et al., 2016; Kress, 2010; Kress and Van Leeuwen, 2001, 2006). When composing a genre, and this is particularly true of genres in digital media, users select the modes from those afforded by the medium (i.e. the material resources, such as the book or the video) and orchestrate them in such a way that they work together to achieve the communicative purpose(s) of the genre (Bateman, 2008; Jewitt, 2016; O'Halloran, 2009). The medium of communication provides the semiotic modes that are available to fulfill certain functions or achieve particular communicative purposes.

The perspective of multimodality enhances the analysis of genre connections in web environments. It allows the analyst to address critically the relations between the various semiotic resources in genres online and the relations between embedded genres and the digital genres or platforms where they are embedded—for example, journal articles online embedding author videos, podcasts or audio slides, or embedded video pitches in crowdfunding project proposals. It also enables us to study how the different genres forming multimodal genre systems (Molle and Prior, 2008; Prior, 2009) create meaning differently, and how form and meaning are remade across modes to achieve the purposes of the genres. We are interested, therefore, in examining how various modes are orchestrated in digital genres to create meaning, i.e. what modes are involved in creating meaning in the genres analyzed and what is communicated by each genre. In addition, we deem it important to investigate relations between modes both within a single genre text with embedded genres and across the different genres forming a network online. Multimodal analysis allows us to examine the interaction between these modes and the distribution of meanings between the different genres. In examining these meanings, it is possible to discern how they are socially situated or, as Paltridge (2012: 171) puts it, how they "are shaped by the norms, rules and social conventions for the genre that are current at the particular time, in the particular context."

The analysis of intersemiosis involves describing the interaction of various semiotic resources in a given context or social site to make meaning (Jewitt et al., 2016; O'Halloran, 2005). In their book on the transcription and analysis of multimodal texts, Baldry and Thibault put forward the "Resource Integration Principle," the notion that the semiotic resources in multimodal texts are "integrated to form a complex whole which cannot be reduced to, or explained in terms of the mere sum of its separate parts" (2006: 18). They complement each other in the meaning creation process and therefore must always be analyzed taking into consideration their interrelations. Intersemiosis has been explored from the approaches of multimodal systemic functional linguistics and multimodal social semiotics, which in turn are based on Halliday's (1978) social semiotic theory. In a multimodal text, various semiotic resources interact to make three different types of meaning or metafunctions: "ideational meaning," to construct our experience of the world ("experiential meaning") and to make logical connections ("logical meaning"); "interpersonal meaning," to take up particular positions, enact relations in social interactions and express attitude and evaluation; and "textual meaning," to organize meaning into a coherent text. O'Halloran et al. (2019) demonstrate how systemic functional multimodal discourse analysis (SF-MDA) can be performed in an analysis of the World Health Organization (WHO) Ebola web page, using the purpose-built software "Analysis Image software" (O'Halloran et al., 2015). A model frequently used to account for the relations between text and image in a multimodal ensemble is that proposed by Martinec and Salway (2005:

342), also based on systemic functional linguistics. This model is used to analyze two types of relations: relations of status—i.e. the relative status of image and text, whether they have an equal or an unequal status, one mode being subordinate to the other— and logico-semantic relations—e.g. elaboration, enhancement, and locution. O'Halloran and colleagues have also developed Multimodal Analysis Video software to perform SF-MDA of dynamic media such as videos (O'Halloran and Lim, 2014; O'Halloran et al., 2015). An example of multimodal analysis of video genres, which shows how intersemiotic relations work to achieve the purposes of the genre, is Lim and Jiang's (2021) analysis of language-visual, language-gestural, and language-visual-gestural intersemiosis in TED talks. These authors state that in these talks a variety of semiotic resources—language, gestures, images on slides—are orchestrated to engage and persuade the audience.

Multimodal texts are designed by selecting and arranging the available semiotic resources to realize particular purposes and to meet the needs of the audience (Bezemer and Kress, 2017). Since the choice and arrangement of semiotic resources is socially motivated, in our analysis of intersemiotic relations in genre networks, we start by considering the purpose(s) of the genre exemplars and then move on to analyze which semiotic modes are used and how they are combined to achieve these purposes (see Bateman, 2014; Van Leeuwen, 2005). This is the starting point from which to interpret the context in which the genre is produced. In order to analyze genres from a multimodal perspective, Van Leeuwen (2005) applies move analysis (Swales, 1990) but makes three assumptions:

(i)   the rhetorical moves (or stages) of a genre may be realized by alternative semiotic resources (not only by text)
(ii) a given move can be multimodal and be realized by multiple semiotic resources simultaneously
(iii) the relations between modes can be of two types: "elaboration," i.e. "content realized in one mode is restated through another mode in specific ways" (e.g. summary, example), and "extension," or when "a move adds new, related content to the content expressed in another mode."

This type of analysis involves looking at what the various modes selected to realize a specific move actually do, that is, when several modes are co-present, how they contribute to the message. Multimodal move analysis has been applied to the study of several digital genres for science communication. Hafner (2018), for instance, describes video methods articles by identifying the moves and steps of the genre, and examines how the written, spoken, and visual resources are orchestrated in the demonstration move, a move where scientists demonstrate research procedures. Similarly, Plastina (2017) conducts move analysis of video abstracts and then explores the relations between the visual and the spoken text in the abstracts.

Studies of (audio)visual rhetoric also provide useful insights for the analysis of intersemiotic relations (see Tseronis and Forceville, 2017). The study of (audio)visual rhetoric addresses how visuals are used to achieve rhetorical purposes, for example to increase the credibility and authority of the producer, engage the audience by appealing to their emotions, or persuade by presenting evidence and arranging it in an optimal way. Tseronis and Forceville (2017: 6) prefer the term "multimodal rhetoric" that includes other modes and focuses on how various semiotic resources, not only visuals, are combined to "address a specific audience in a certain rhetorical situation." This perspective expands models of discourse and genre analysis that have until now been applied only to the verbal mode. An example is De Groot et al.'s (2016) analysis of visual metadiscourse in the annual reports of companies. These authors use a model of visual metadiscourse that extends the textual metadiscourse framework developed by Hyland and Tse (2004) to include Kress and Van Leeuwen's (2006) theory on multimodality. Metadiscourse is related to Halliday's textual and interpersonal metafunctions, used by writers to organize the text, guide readers' interpretations, engage the audience, and create reader involvement. Images also have an interpersonal function (Kress and Van Leeuwen, 2006), and can be used for similar purposes as textual metadiscourse.

In addition to the analysis of intersemiosis, the second type of multimodal analysis that we are concerned with is the analysis of relations among multimodal texts (as exemplars of genres), where one of the texts results from the transformation of the other. We follow previous studies that go beyond the analysis of intratextual semiotic relations and focus on semiotic relations between texts/genres (Bezemer and Kress, 2017; Bezemer and Mavers, 2011; Iedema, 2003; Prior and Hengst, 2010). In recent decades, several concepts (sometimes overlapping) have been used to theorize and explain these relations: "remediation" (Bolter and Grusin, 1999), "media transformation" (Elleström, 2010), "resemiotization" (Iedema, 2003), "transmodal redesign" (Mavers, 2011), "re-entextualization" (Blommaert, 2005; Gimenez et al., 2020), "recontextualization" (Bernstein, 1996; Bezemer and Kress, 2008), "multimodal recontextualization" (Androutsopoulos and Tereick, 2016), and "transduction" (Bezemer and Kress, 2008).

For the analysis of resemiotization processes, we use the concepts of remediation and recontextualization, following Bezemer and Kress's (2008, 2017) definitions of these concepts. As explained in Chapter 1, remediation is the movement of meaning material from medium to medium, harnessing the modal affordances of the new medium (Bezemer and Kress, 2008). This is an important concept in the discussion of genres for science communication online, since many of these genres have antecedents in non-digital media and are a result of the migration of these pre-digital genres to the digital environment. Remediated genres may establish connections with other genres, which were not possible in other media. The printed journal, for instance, is remediated into enhanced publications that exploit the affordances of the medium to integrate dynamic components—e.g. video, sound files—and interactive components and hyperlinks (Blevins et al., 2015). This combination of verbal and visual resources in the remediation of scientific knowledge into digital genres enables scientists to engage in showing rather than telling. Blevins et al. (2015: online) claim that "turning a print-based piece into multimodal communication involves considering the ways in which communication modes interact and complement one another." These authors put forward a guiding principle to consider in this process: implementing design, taking into account both the resources available in the specific context where communication will take place and the purpose and audience of the genre. Since different modes are more appropriate for conveying different types of meaning, producers of multimodal scholarly texts have to consider multimodal affordances and choose the best modes for their particular purposes, audiences, and platforms. For Blevins et al. (2015: online), the remediation of scholarly printed texts into online texts is a process of re-design—a concept similar to re-entextualization—which involves asking questions such as: How is the content best re-organized in terms of space? How is the content most effectively re-presented through linear or non-linear arrangement? How might the content be re-delivered in ways that speak to multiple audiences? How might the content be re-conceptualized aurally, visually, and via animation? Since these are questions considered by rhetors when redesigning texts, they should also guide the analyst when exploring how texts are re-designed in a different mode.

A related concept is recontextualization, already discussed in Chapter 1. A definition of this concept which is particularly relevant in the present study of multimodal digital texts is that put forward by Bezemer and Kress (2008: 184), because it is based on the assumption that meaning is made through the interaction of modes. They define recontextualization as follows:

> moving meaning material from one context with its social organization of participants and its modal ensembles to another, with its different social organization and modal ensembles. Meaning material always has a semiotic realization, so recontextualization involves the re-presentation of the meaning materials in a manner apt for the new context in the light of the available modal resources.

According to Bezemer and Kress's (2008: 184–186), recontextualization involves four rhetorical principles: selection (of meaning material to be moved that will be relevant in the new context, and of modal resources which are available in the new context and are appropriate for the audience of this context); arrangement (of the meaning materials in a way that is best for the new context); foregrounding (of the elements that are particularly significant in the new context); and social reposition (or reconstruction of social relations between the recontextualizers and the audience of the new context). All these questions are related to the concepts of transduction and transformation (Bezemer and Kress, 2008). Transduction refers to the movement of meaning across modes using the semiotic resources available in a given context—e.g. remaking a written text as a diagram—and transformation refers to changes within a mode.

The principles proposed by Bezemer and Kress (2008) are clearly illustrated in Fryer's (2016) analysis of the process of creation and online recontextualization of a figure, originally published in print and online in a medical research article. Since its publication in the journal, the figure has been reused and repurposed in different genres—e.g. in other research articles, in weblogs, on various websites, on YouTube videos. Fryer (2016) shows that the rhetors' decisions regarding the selection, arrangement, and foregrounding of meaning material as the figure is reused in different contexts are determined both by the modal resources available in the new context, by the rhetors' own interests and motivations, and by the perceived expectations of the audience. Androutsopoulos and Tereick (2016) also illustrate multimodal recontextualization practices in their study of YouTube. As in the case of the figure in Fryer's study, one important feature of YouTube content is detachability, which facilitates intermedia circulation. Two multimodal recontextualization practices through which users engage with others' video material are embedding and remixing. In the case of YouTube, embedding consists of the uploading of an existing video from other sources—e.g. TV programs—and giving it a new title providing a short description. Remixing is the modification or manipulation of a single video or the blending of excerpts from several videos into a new video (Androutsopoulos and Tereick, 2016). In both cases, the videos provided by other users are re-entextualized to convey new meanings and achieve new purposes.

In the case study research, we analyze how meaning is transformed in the various genres forming an assemblage; that is, we explore strategies of multimodal recontextualization. We intend to answer the following questions: What elements (i.e. language, contents, style) of the source text are selected to be moved to the new context? Are these elements recontextualized by using the same modes as in the source text or different modes? What recontextualization strategies are used, i.e. how are these elements transformed? To answer the last questions, Van Leeuwen (2008) analyzes the kind of transformations that take place in the process of recontextualization and identifies four strategies of recontextualization, which can take different forms and thus result in different types of meaning transformation. These strategies are substitution of elements—e.g. through particularizing and generalization—deletion, rearrangement, and addition—e.g. repetitions, reactions, evaluation.

The first analytical step in tracing recontextualization in the case studies consists of a close intertextual comparison to identify the "tangible traces"

(Bray, 2019: 165) of the source text in the text that recontextualizes it, that is, the meaning material that has been moved to the new text. The difference in the meaning materials in both texts and in how they are realized linguistically and visually provides the basis for the identification of multimodal recontextualization strategies, e.g. rearrangement, addition, or foregrounding (see Bezemer and Kress, 2008; Luzón, 2013; Van Leeuwen, 2008). The identification and coding of these strategies can be facilitated by software for qualitative analysis such as ATLAS.ti or NVivo so that the analysis is empirically grounded.

Genre analysis and hypermodality

Hypermodality, or "the conflation of multimodality and hypertextuality" (Lemke, 2002: 301), is also a fundamental feature for understanding and analyzing genre relations in digital media environments. Lemke (2002: 300) uses the term *hypermodality* to refer to "new interactions of word-, image-, and sound-based meanings in hypermedia, i.e. in semiotic artifacts in which signifiers on different scales of syntagmatic organization are linked in complex networks or webs," thus emphasizing that in digital genres multimodality and hypertextuality—i.e. the connection of electronic texts through hyperlinks—cannot be analyzed independently. Lemke (2006: 6) also defines the web as a "hyper-textual or hypermedia medium," where links can connect elements in different media and modalities. As this author puts it, "[a] trajectory across links within a website may already carry us across different genres and different media using different modalities" (Lemke, 2006: 6).

Several authors have drawn attention to the intertextual and heteroglossic nature of digital genres (Androutsopoulos, 2011; Casper, 2016; Lemke, 2002). Two affordances of digital media that are key in facilitating the creation of relations between genres within a single website and across websites or platforms are hyperlinking, which enables the creation of hypertext, and modularity, which makes it possible to create digital texts composed of autonomous modules, each of them with a particular function. These affordances therefore facilitate genre embedding and genre hybridity.

Linking enables the writer to compose multisequential texts and offers readers many possible trajectories (Lemke, 2002), thus meeting the interests and information needs of diverse audiences. Hypertext allows the reader to access connected texts easily and to move through non-linear unbounded spaces. As noted in Chapter 1, readers decide what content to access and what content to skip, as well as the order of access to the materials, depending on their own goals and interests. However, on the web the reader's agency somehow fuses with the rhetor's agency, who deliberately creates connections between texts through hyperlinking or embedding. Hyperlinks connect texts to prior texts for multiple purposes. As text producers, rhetors can use hypertext links to organize a website and provide the reader with access to different types of information on the website (internal links) or to link to other texts on the Internet for several purposes, such as supporting the rhetor's claims or helping readers access writers' research output on other platforms (text external links). Since, by linking to other texts, text producers establish semantic and rhetorical relations between the texts that are linked and create relations between genres, it is important to examine these relations.

Understanding hypertextuality in digital genres for science communication requires analyzing the types of links in these genres, the elements they link—i.e. the source and target texts of the link—and their function (Bar-Ilan, 2005; Luzón, 2009, 2017; Thelwall, 2003). For instance, the framework developed by Bar-Illan (2005) to analyze the use of links between academic institutions classified these links taking into account several aspects, such as the source and the target page (a concept similar to genre or part-genre, e.g. personal homepage, journal instructions, call for papers), the placement of the link (e.g. the link may be part of a list, embedded in the body of the text, or appearing in the sidebar/menu/logo), and the relation between the source and the target (e.g. the source expands/provides information/provides technical detail/summarizes/is part of the target). In her analysis of research group blogs, Kim (2000) analyzed links from scholarly electronic papers and found that they were used for purposes such as providing additional or background information, illustrating a point, publicizing an information source, giving credit to an author or institution, or supplying access to a document. Similarly, in her analysis of links in academic blogs, Luzón (2009) found that bloggers linked very frequently to other posts within their own blog or to other pages on the Internet referring to their research—e.g. pages with the blogger's books, research, interviews, or courses—but also frequently to posts by other bloggers or to other researchers' output. These types of links within the blogosphere are used by researchers to engage in hypertext conversations for the collaborative construction of knowledge. In a later study on blogs by research groups, Luzón (2017) showed that when composing their posts, writers incorporate a variety of genres by linking (e.g. scholars' homepages, online abstracts, online papers, posts in other blogs, project homepages, lectures on YouTube), or embedding (e.g. graphical abstracts, calls for papers, Slideshare) and interconnect these genres to achieve several objectives, such as publicizing the group's research and activities or strengthening social links within their community. Links in blogs are also used to address the information needs of different audiences (Luzón, 2013). For example, in posts where bloggers discuss published papers, links can be used to provide more detailed information for experts (e.g. links to research papers) or to help the interested public understand disciplinary concepts (e.g. links to a Wikipedia entry). These studies reveal that writers of digital genres use links to other genres strategically to achieve specific purposes. Therefore, it is relevant to study patterns of genre relations created through links, specifically whether some genres tend to be hyperlinked with others and why.

In the case study research, we take into account the hypertextuality of genre networks to ascertain how and why the contents of a genre are enhanced through hyperlinks. We also critically discuss the effects of the hypertextual affordances in genre networks and how incorporating other rhetors' voices and perspectives supplements the interpretation of each genre. In tracking genre interrelations based on hypertextual links, we show how and why the textual boundaries of these connected genres are blurred, as hypertextuality offers the reader the possibility of accessing related contents conveyed from new and diverse perspectives and viewpoints. Such non-closure, also inherent in intertextuality, sheds important light on our understanding of intersemiotic relations in a broader context—a context that subsumes all the specific contexts associated with each genre of the network. It should nonetheless be made clear here that we only discuss aspects of hypertextuality to highlight or reinforce the existing connections between or among genres that we have traced at the level of semantics, grammar, and discourse pragmatics.

In addition to taking into account the function of hyperlinks, we also comment on links to discover how a text travels from one digital space to another and is repurposed in different genres. We assume that one text may be re-entextualized into multiple texts, rather than one single text which, echoing Gimenez et al. (2020), captures a more accurate picture of how in the connected genre exemplars of a network, contents are reorganized, refocused and their contents reinterpreted. For instance, as we will see in Chapter 4, a research article may be re-entextualized through processes of discoursal uptake in a blog entry, an online news item, or a tweet, which may include links to the original research article.

In this chapter we have advocated the exploration of interrelated genres online from multiple analytical perspectives. Echoing Bhatia (2004: 155), our main argument for proposing this methodology is that our focus of analysis, networks of connected genres online, entails complex discourse construction and knowledge transformation processes. Accordingly, the chapter has described how the integrated analytical approach can better capture how knowledge is (re-)entextualized in genre networks and help the analyst to identify more easily the contexts—socio-historical, social, cultural, institutional, and economic—that specifically shape each genre and motivate the connections between or among genres. In the following chapter we apply this analytical approach to a selection of case studies.

4 Case study research

Case study 1. High-stakes science

In this case study we explore processes of discoursal uptake and recontextualization within a network of genres that represents three types of discourse—institutional, scientific, and para-scientific. These discourses involve text creation by different science stakeholders with diverse perspectives on the same topic: the rising sea level resulting from climate change. We critically discuss the construction of meanings around this topic by examining the different textual layers and identifying the intersecting meanings across the texts. To do so, we compare the following aspects in the genre exemplars of this network: (i) the degree of semantic and syntactic overlap, i.e. shared lexical and grammatical features, (ii) the presence of intertextual links, and (iii) the linguistic realizations of authorial positioning and pragmatic stance towards the topic of the sea level rise. We aim to understand how the different social actors, as rhetors, project their perspectives and views about climate change onto their texts and position themselves towards this topic in various ways. In doing so, we wish to show how this network offers a polyphony of voices and a wealth of viewpoints.

The focus of this first case study is on the re-entextualization of a web-based text summarizing the Paris Agreement reached at the United Nations Framework Convention on Climate Change,[1] available on the UN website and YouTube. This text is recontextualized in several hyperlinked texts: a *Nature Communications* original research article[2] dealing with the topic of sea level rise, its associated editorial[3] together with a short text online introducing the article, and a blog post written by a freelance science writer[4] that mentions the article. The web-based text of the Paris Agreement totals 746 words and the count of the transcript of its embedded video (lasting 1:39 mins) is 191 words. The *Nature Communications* editorial of the Climate Change collection is 1,020 words and the editor's short text introducing the research article is 49 words long. The article and its abstract account for 5,311 and 177 words, respectively. Finally, the blog post written by the science journalist is 2,937 words. Figure 4.1.1 provides an overview of the genre network under analysis in this case study.

DOI: 10.4324/9781003147732-4

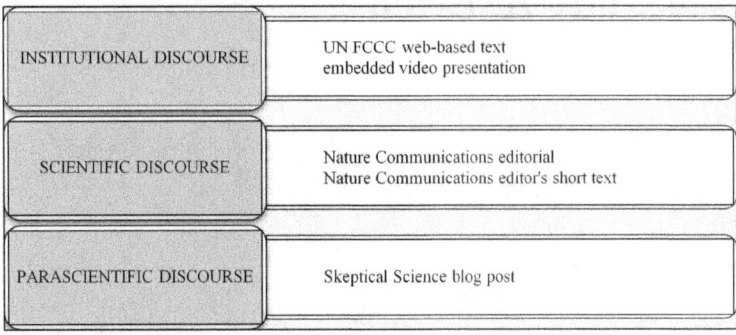

Figure 4.1.1 Discourses and genre types represented in the network.

We use corpus tools and descriptive statistics to compare the different text varieties and track interrelations among them, by this means seeking to understand the meaning-making processes involved in this network. The most informative measure is STTR. As Scott (2008) explains, this measure is used to compare type/token ratios across texts of differing lengths, which is the case here. Because the network includes texts belonging to different types of discourse, we compare frequency lists and carry out a concordance analysis using simple searches of "climate change" and "sea level." We briefly address how multimodality complements the semantic meanings conveyed in some of the texts in this network. We also use measures of lexical and syntactic complexity to compare the conceptual depth of the texts. In addition, as stated previously, we pay attention to linguistic markers of modality and authorial stance, following White (2003), with a view to identifying different views towards the effects of climate change and the underlying ideologies in the three types of discourse represented in the network.

We deem it of interest to examine how the institutional discourse of the Paris Agreement presented in a web-based environment is taken up by the editor of a high-stakes journal, *Nature Communications*, and by scientists who, representing the voice of scientific authority, present their expert views on the natural phenomenon under discussion and by this means re-entextualize the source text in texts representing scientific discourse. We also analyze how the expert discourse of the original article is subsequently recontextualized and adapted to para-scientific discourse, and how the scientists' research-informed views are refocused by the science journalist in a blog post.

Shared semantic meanings and intertextual links

To trace the semantic interrelations of the genres forming this network, we first look at their lexical and syntactic profiles and afterwards at the intertextual connections among them. As shown in Table 4.1.1, in this case study

Table 4.1.1 Overall descriptive statistics of the network

Texts	Type/token ratio (TTR)	No. sentences	Mean (in words)
The Paris Agreement	42.424244	43	16.883722
The Paris Agreement embedded video	61.62162	12	15.416667
Editorial *Nature Communication*	46.693386	32	31.1875
Editor's short text on authors article	83.67347	2	24.5
Research article abstract	58.490566	7	22.714285
Original research article	18.399019	213	22.990612
Science blog post	35.729462	116	24.344824

the TTR and STTR measures do not clearly indicate the amount of lexical diversity in each text, as they vary considerably in length. We can only compare the editorial and the blog, which have similar word lengths, and then we obtain very similar STTRs (46.099998 and 46.150002, respectively). The STTR of the *Nature* article (33.32) shows a comparatively lower lexical diversity. Other measures, such as the mean (in words) of the texts give clearer hints about them at the sentence level. The Paris Agreement and the transcript of its embedded video score lower average means in words per sentence (16.883722 and 15.416667, respectively) compared to those of the remaining texts, as shown in Table 4.1.1, suggesting greater syntactic complexity than that of the text on the UN website and its embedded video introduction.

The comparison of the frequency lists of these texts reveals a range of semantic meaning interrelations within the genre network. In the Paris Agreement, which is the source text that is later re-entextualized in the editorial and the article, the top content words are "countries," "climate," "change," "emission," and "development." The highest frequency collocate "climate change," occurring 104.948 times per 10,000 words, brings to the surface an evaluation-laden co-text expressing the commitment of the UN to "undertake ambitious efforts to combat climate change" multilaterally, that is involving all countries worldwide. As stated on the UN web page:

> The Paris Agreement is a landmark in the multilateral climate change process because, for the first time, a binding agreement brings all nations into a common cause to undertake ambitious efforts to combat climate change and adapt to its effects.

Meaning making in the UN web page on the Paris Agreement is also conveyed by the audiovisual/verbal message in its accompanying embedded video. The function of the video is to introduce a synthesis report of the Paris Agreement. The video uses three modes, combining oral narrative with visual elements (e.g. colors, icons) and written messages. Through these modes the initial shot of the video reinforces the idea that this is a "legally binding" and "international" agreement, aspects that the web-based text expresses using the words "legal" and "multilateral." The subsequent shots visually reinforce the details given by the oral narrative ("The Paris Agreement is a legally binding international treaty on climate change to limit global warming to well below 2"). Afterwards, the visual information summarizes the main goal of the agreement and the specific action taken to address climate change, namely "reduce greenhouse gas emissions." These shots involve simple meaning representations: a drawing on the left of the screen and the short message on the right. Towards the end of the video, the oral narrative and the visual and verbal messages foreground the anticipated effectiveness of institutional commitment and the undertakings of nations to fulfill the goal of the agreement by setting a timetable to track progress of the actions implemented to combat climate change and to do so in a transparent way. The oral narrative ("Starting in 2024, countries report transparently on actions taken. Collective progress under the Paris Agreement will be assessed through a global stocktake") is accompanied by several shots with an icon that reinforces the "globality" of the action.

The texts recontextualizing the original web-based text exhibit greater lexical diversity and syntactic complexity, which indicate greater conceptual depth. These language features are used by the rhetors, the *Nature Communications* series editor and the article authors, to bring in new perspectives—namely, those of the "experts" or "specialists" in climate change and sea level rise. The editorial explicitly draws on the intertextual reference to the Paris Agreement to present the problem—i.e. "rising seas"—as an effect of climate change, as deduced from the semantics of the immediate co-text of the content words "sea," "level," and "rise" (occurring with a frequency of more than 10 times in this text). In this text, the collocate "climate change" occurs with a relative frequency of 31.120 per 10,000 words. The KWIC concordance lines of the collocate "climate change" indicate that the editor takes an overt evaluative stance when envisioning the future effects of it, as in the statement "[...] prioritize efforts to avoid dangerous climate change" or even in a single, but syntactically complex, sentence built upon parallel structures—two non-finite *-ing* clauses, the second embedded in the first—that highlight the predicted socioeconomic consequences through an intertextual reference ("one study forecasts that ...").

> Under an extreme sea-level rise scenario, one study forecasts that as many as 13.1 million US citizens could become climate change refugees, triggering a wave of mass migration away from the coastal zone, resulting in a heavy socioeconomic burden on the infrastructure of inland cities.

Instances of manifest intertextuality can be found in the short summary that the editor provides to introduce the original research article and to foreground causes ("emissions") and consequences ("sea level rise") while also acknowledging "uncertainty." The editor packages all these meanings in a single syntactically complex sentence placing the new information in a *that*-complement clause explicitly referring to the Paris Agreement and the uncertainty of the future:

> Here the authors show that under the Paris Agreement, emissions in the next decades have a strong influence on the amount of sea level rise in the centuries to come, with the uncertainty dominated by ice-sheet contributions.

The research article re-entextualizes the discourse of the UN Paris Agreement by narrowing down the issue of global climate change and by focusing on the phenomenon of sea level rise. These are in fact the top three highly frequent content words. The first two, "sea" and "level," are especially frequent in this text, occurring more than 100 times. Along with the word "emissions," these three content words are also the highest frequency content words in the article's associated abstract. Keyness analysis further confirms that the article is a domain-specific text. Compared with a reference corpus of general English (the BROWN corpus), the words with the highest keyness are "sea" (with a keyness value of 837.359924), "level" (733.983093), "rise" (626.705566), and "emission" (609.372803). Table 4.1.2 shows the 50 words with the highest keyness in this text. Wordgrams (f > 5), retrieved with Fletcher's (2002–2007) software, also reveal features of densification, as deduced from the presence of nominal phrase structures with nominal premodification such as "for net-zero GHG scenarios," "the Antarctic ice sheet," "Greenland surface mass balance," and "net-zero co# scenarios and," among others.

Table 4.1.2 Top 50 words with highest keyness (cut-off point > 100.000)

N	Key word	Keyness	N	Key word	Keyness
1	sea	837.359924	26	mean	107.635452
2	level	733.983093	27	emission	107.211571
3	scenarios	651.443787	28	percentile	104.976906
4	rise	626.705566	29	instability	104.976906
5	emission	609.372803	30	contribution	103.449318
6	zero	542.487854	31	model	97.5566864
7	ice	541.509216	32	temperatures	95.4555664
8	net	515.336609	33	overshoot	94.4775848
9	global	472.682526	34	simulation	94.4775848
10	GHG	388.596161	35	loss	94.056282
11	Greenland	304.533142	36	agreement	87.3713379
12	median	283.521027	37	sensitivity	85.5508347
13	temperature	267.787598	38	parametrization	83.9786224
14	climate	256.543335	39	estimates	80.4001465
15	scenario	252.005615	40	component	78.8524399

(Continued)

Table 4.1.2 Continued

N	Key word	Keyness	N	Key word	Keyness
16	Antarctic	188.984573	41	mass	78.8302002
17	warming	171.904343	42	estimate	78.4044724
18	sheet	170.021484	43	calibrated	73.4800262
19	discharge	156.149918	44	ocean	72.9431534
20	feedback	136.477066	45	balance	71.4216461
21	glaziers	125.976646	46	peak	70.4988327
22	stabilization	115.476593	47	observations	70.2162781
23	melt	115.476593	48	ensemble	67.0678329
24	response	112.080528	49	reductions	62.9817886
25	Paris	108.396355	50	fossil	62.9817886

Building upon this lexicogrammatical profile, the semantics of the article revolves around "ice," "scenarios," "zero," "emissions," and "temperature," each of them occurring over 40 times. The collocate "climate change" is used only in the first sentence of the abstract, but the utterance containing it ("Sea level rise is a major consequence of climate change that will continue long after …") is forceful enough to emphasize the scope of the problem. In the research article, the collocate "climate change" occurs 20.227 times per 10,000 words, and special emphasis is given to "sea level rise" that also occurs in extended collocations such as "sea-level rise in 2300," "median sea-level rise in," and "long term sea-level rise" (f > 5 times). The collocate "sea level" occurs 12.136 times per 10,000 words. The semantics of its immediate context places the emphasis on the fact-based problem, formulated in an extraposed *that*-complement clause that explicitly cites the Paris Agreement, as shown in the extract (our emphasis added).

> We find that <u>sea level continues to rise</u> in almost all cases throughout 2300, with sea-level stabilization possible but not probable under declining global-mean temperatures. […] Furthermore, we find that a delay of global peak emissions by 5 years in scenarios compatible with the Paris Agreement results in around 20 cm of **additional median <u>sea-level</u> rise** in 2300.

A possible solution, based on facts, is formulated once again through an extraposed *that*-complement clause that gives greater weight to what is expressed in this clause.

> […] net-zero GHG scenarios indicates that <u>sea level</u> *stabilization is possible* under declining global mean temperatures.

Finally, the science blog borrows part of the semantics of the *Nature* article, yet in a slight different way. The top three highly frequent words, "climate," "sea," and "ice" (at a cut-off frequency > 20) reveal a somewhat different lexical profile compared to that of its pretext. The collocate "climate change," occurring 13.947 times per 10,000 words, is less frequent than the collocate "sea level" (41.841). This was not so in the article, which suggests different repurposing strategies in the process of knowledge transformation within this network.

Authorial voice

Here we look at the construction of arguments in relation to the overall rhetorical organization of the texts, and later address aspects of authorial voice as traced through the expression of modality meanings towards the topic of climate change. Specifically, we analyze the use of modal verbs conveying intrinsic and extrinsic modality (Biber et al., 1999), also conceptualized as deontic and epistemic modality, respectively (Facchinetti and Palmer, 2004). These verbs are markers of stance and attitude that signal different clines of authorial positioning towards the utterances expressed in the texts. Table 4.1.3 compares the frequency of modal verbs (based on the raw count of modals included in the frequency word lists (>5)). The comparison shows a wider range of modals in the article and in the blog than in the short text, possibly because the former are much longer texts. The LL values calculated with the log likelihood and effect size calculator at the University of Lancaster[5] further reveal statistical significance in the use of "will," especially frequent in the UN web-based text compared with the *Nature Communications* article, and the use of "could," that proves to be especially frequent in the editorial compared to the blog. These data set the basis for the subsequent commentary of aspects of authorial positioning and their discourse pragmatics effects.

In the UN web-based text, the information is visually arranged in two parts that are clearly identified with headings ("What is the Paris Agreement" and "How does it work?"). In this text the modal verb "will" is used in assertive statements that reaffirm the commitment of participating countries to make the five-year process of action implementation transparent and assuming their willingness to accept responsibility for the actions they take. Using this certainty modal, the text reinforces the UN interest or intention (our emphasis added).

Table 4.1.3 Comparison of highly frequent modals (normalized frequency per 10,000 words)

UN *web page text*	Nature *editorial*	Nature *article*		Science *blog*	
will 80.4289544	could 68.627451	can	30.1261533	will	30.6435138
		will	15.0630766	may	23.8338441
		may	11.2973075	could	20.4290092
		cannot	9.4144229	can	17.0241743

> Under EFT [enhanced transparency framework], starting in 2024, countries <u>will report</u> transparently on actions taken and progress in climate change mitigation, adaptation measures and support provided or received. [...] The information gathered through the ETF <u>will feed</u> into the Global stocktake which <u>will assess</u> the collective progress towards the long-term climate goals. This <u>will lead</u> to recommendations for countries to set more ambitious plans in the next round.

In the *Nature Communications* article, the expression of certainty through "will" also introduces bare assertions. However, these are predictions based on compelling scientific evidence accompanied by overt criticism towards the, so far—in the authors' view—ineffective actions taken by countries to mitigate climate change. The personal voice of the "experts" is not overtly expressed at a textual level. Rather, it is formulated through inanimate subjects. Epistemic meanings of possibility and prediction (or extrinsic modality) are found in the discussion section of the article. At the start of this rhetorical section, the intertextual reference to the Paris Agreement brings to the fore the authors' personal understanding—this time formulated by the pronoun "we"—of the effectiveness of the goals sought by the Paris Agreement to be achieved in two centuries' time. However, later in the text, oblique forms of this pronoun ("our results," "our analysis") are used to question in a more neutral tone the effectiveness of the agreement ("vital [...] but insufficient"), as stated in the closing sentence of the following extract.

> In this work, we link stylized emission scenarios that reflect the Paris Agreement goals to sea-level rise in the year 2300. Our results indicate that sea-level rise <u>will continue</u> until and beyond 2300 even for scenarios that reach net-zero GHG emissions in the second half of the 21st century. [...] By combining climate and sea-level uncertainties, our analysis reveals a persistent risk of high sea-level rise even under pathways in line with the Paris Agreement.

As the discussion unfolds, the authors offer their opinion by taking a different stance in order to let scientific facts speak for themselves. This is done through inanimate subjects, with which the authors implicitly reiterate that the meanings conveyed in the statements regarding the potential negative effects of climate change in the future are to be taken as the bare truth. In other words, the authors' scientifically proven predictions lead them to the assumption (and indirect criticism) of a "slow response" or "delayed near-term mitigation action." Their informed opinion to mitigate climate change effects through the "vital and important" actions the scientists propose ("Early peaking and stringent emissions reductions") is based on objective facts, namely their study results. Objectively formulated by inanimate subjects, meaning making revolves around attested facts and not personal opinions.

Due to the slow response to climate warming, <u>sea level will continue to rise</u> after temperatures have stabilized. <u>Positive rates of sea-level rise will likely persist</u> through 2300 even under net-zero GHG emissions [...] Early peaking and stringent emissions reductions thereafter are vital and important to reduce the risk of low-probability high-end sea-level rise, yet insufficient to stop global sea-level rise by 2300. Delayed near-term mitigation action in the next decades <u>will leave</u> a substantial legacy for long-term sea-level rise.

Different perspectives or viewpoints are expressed in the *Nature Communications* editorial and the blog. By tracing the use of "could" as a modal marker expressing deduction, we find that the editor makes inferences based on the scientific facts reported in the article, specifically on the implications of the study results. In all the statements where the modal "could" is used, the editor repurposes the article contents by highlighting the causes rather than the effects of climate change. The article authors also do so, but not in such an overt manner. A randomly selected concordance line easily illustrates the way this modal marks a logical possibility: "Variations in these residual non-CO_2 GHGs could result in higher or lower rates of temperature decline."

In the science blog, though, the expression of personal meanings modalized by the verb "could" builds a different type of deductive argumentation, laying bare the rhetor's overt commitment towards what is expressed in the text. The textual information is structured into a sequence of paragraphs. Each of them answers a rhetorical question that acts as the heading of each paragraph. These rhetorical questions are used by the science journalist to raise several points and build central arguments throughout the text. The blog post does not contain any manifest intertextual reference to the Paris Agreement, although in its initial sentence the journalist does refer to climate change, inviting readers to imagine a future scenario and by this means warning the reader of the immediacy of climate change effects. Later, the journalist draws on arguments from scientific sources to construct an objective and factual account of sea level rise. The construction of the arguments is supported by abundant intertextual references to other scientific works, including an explicit attribution to the *Nature Communications* article ("According to Matthias Mengel ...") and two short direct quotes from the article text. Yet, notwithstanding the explicit reference to the article, the blog post recontextualizes its contents and refocuses them to provide a personal perspective, directing attention to the two assumptions ("big uncertainties") resulting from the explanations for the sea level rise provided by the sources cited in the blog, namely "this assumes" and "this future also assumes." The hypothetical situation in future time ("our descendants") is expressed by means of "could" which, along with the use of the first-person pronoun "we" and the oblique form "our," create a persuasive appeal to the reader. This is shown in the following stretch of discourse that conjectures about the issue of sea level rise in a very persuasive manner.

> "Negative emissions" (actively sucking CO_2 out of the air) could slowly reduce global temperatures and stabilize many sources of sea level rise during the 22nd century. According to Matthias Mengel of the Potsdam Institute for Climate Impact Research and colleagues, falling CO_2 <u>would</u> eventually <u>allow</u> Antarctica to begin accumulating ice, so sea levels <u>would begin</u> to fall again, three centuries into the future. But this assumes negative emissions technologies can be deployed massively by the 2030s, a scenario with "limited realistic potential." Every five-year delay <u>could</u> commit **our** descendants to an extra 1 meter (3 feet) of sea level rise by 2300. Avoiding this future also assumes **we** don't trigger widespread ice sheet collapse in the meantime.

As shown here, in the process of recontextualizing expert knowledge, the focus of this text is placed neither on the resultant natural effects ("sea level rise") nor on its causes, but on the hypothetical future impact on people's lives (again, "our descendants") that the current uncertainties may have. In short, the journalist redirects his critical reflection to what the scientists actually report in their article.

Interpretation

In this genre network we have analyzed several typified—"stabilized-for-now" (Schryer, 1994: 108)—genre exemplars which, as can be deduced from their main lexical and grammatical features, are clearly associated with scientific, expert-to-expert discourse. These are the editorial, the editor's short text, the original research article, and its related abstract. These texts represent high-stakes scientific discourse in a high-impact-factor journal, hence the title of this case study. Our point of departure for the analysis of intersemiotic relations has been the article, through which we have traced a number of intertextual links with genres that belong to other types of discourse. First, we have considered a web-based text and its associated multimedia element (the UN text and its embedded video) that represent institutional discourse. Second, we have analyzed the process of transformation of the content and perspectives of the article in an emerging digital genre, the science blog. This genre exemplar, written by a science journalist and intended to reach a wide audience, aims to represent para-scientific discourse. Further, the case has illustrated, paraphrasing Swales (1990), how the rationale behind each genre establishes constraints on allowable contributions in terms of their content, positioning, and form. The analysis of the lexicogrammatical profile of the texts and the exploration of aspects of intertextuality, textual rhetoric, and discourse pragmatics offer a better understanding of the interactions among genres resulting from processes of discoursal uptake and recontextualization in this network.

Turning first to text-internal features, the case study has shown that the semantic profiles of the different genres forming this network, identified by means of corpus tools, enable the analyst to identify a number of semantic interrelationships among the genres. The results of the corpus analysis have shown that the different genres forming this network do share some semantic meanings. It is precisely this shared semantic layer that evinces the processes of transformation, recontextualization—or re-entextualization, if we use Gimenez et al.'s (2020) term. Notwithstanding this, each genre (re-)entextualizes or creates meanings in a particular way, either by summarizing or by compressing these meanings, as the web-based text and its embedded video do, or by elaborating explanations and relying upon scientific data, as the research article does, or by clarifying meanings, as the editorial does. Acting as a para-text, the editorial recontextualizes the article contents, foregrounding causes. In contrast, the blog re-entextualizes such meanings by putting the emphasis on the implications. The rhetor, the science journalist, does this using persuasive rhetoric.

Notwithstanding the fact that they share certain semantic meanings, as observed in the comparison of frequency lists and recurring collocates and by the fact that all the texts revolve around the topic of climate change, these different texts have been characterized through distinct grammatical and syntactic features. As commented earlier, some of the texts in this network were built upon simple syntactic patterns and showed low lexical density and, hence, a low level of conceptual depth. Other texts, however, contained grammar and discourse style features that are especially frequent in academic prose. The use of such language forms provides a minute account of scientific facts as well as detailed explanations based on empirical grounds. Measures such as keyness and the presence of certain extended collocations have also revealed the presence of "economy/phrasal grammatical features" (Biber and Gray, 2016: 322, see also Biber and Gray, 2010), such as compressed noun phrase structures with pre-modifying nouns, which indicate that these texts can be categorized as domain-specific and informational writing.

Other text-internal features such as the presence of recurring word combinations revolving around the topic of climate change—the central phenomenon—have further shown that meaning making in the research article is strategically articulated to cover all the key aspects of the central phenomenon: the context of the problem ("Antarctic ice sheet"), the intervening condition ("the Paris Agreement"), the causal conditions ("sea-level rise"), and the expected outcomes ("net-zero GHG scenarios"). It is also worth noting that at the level of discourse semantics, whereas the words with the highest keyness are precisely those that discuss the central phenomenon in this text ("sea level rise"), the main emphasis of the *Nature* series editor's editorial and short text was refocused towards "sea level rise" in relation to climate and global warming, the latter echoing the topic addressed by the UN web-based text in very broad terms. Moreover, as also noted above, by complementing these preferred "focuses," the post places the attention on both causal conditions: the natural phenomenon, the sea level rise, and the intervening condition, the Paris Agreement. Tracing meaning-making processes can thus help the analyst reconstruct how the scientists' construction, understanding, and interpretation of the natural phenomenon does not stand by itself in the network. Rather, meanings are co-constructed, resulting from the transformation of the contents of the source texts—or pretexts—and at the same time from the rhetors' personal interpretation and perspective towards such contents and views.

With this case study we have shown how knowledge co-construction can be traced at the level of discourse by specifically looking at the intertextual links between or among the texts forming the network. In tracing features of intertextuality, we have been able to describe the re-entextualization of a succinct text into more conceptually elaborate texts, based on evidential facts, whose contents are subsequently repurposed in a para-text (the editor's short text) that mediates the knowledge base required to understand the article online as well as in a persuasively appealing text (the blog post written by the science journalist). The features of manifest intertextuality exhibited in these genre exemplars—at times in the form of mentions, at other times explicit attributions and direct quotations—reveal semantic interrelations among the texts. In addition, they foreground the complementariness of perspectives and viewpoints towards climate change that a network of interrelated genres can contain.

This wealth of perspectives surfaces when examining the degree of authorial positioning that the different rhetors take towards the utterances they appropriate from the texts they cite. The authors of the article accept the propositions expressed in the Paris Agreement web-based text only to some extent, as they also partially disagree with the source text's prediction about the proposed ways to tackle climate change. This is also evident in the blog post. In taking up arguments from scientific discourse, the rhetor repurposes the contents and constructs arguments that appeal to the validity and truthfulness of scientific authority. While the "aboutness" of the network is consistent across its constituent genres and semantic overlapping thus becomes a clear textual indication of the connections between the genres forming this network, the blogger adds a further perspective, the scientifically based predictions of an uncertain future are contested ("Wait, there is hope!").

In this case study we have not commented on visuals in much depth since other case studies will place greater focus on aspects of multimodality. We have only sketched out how a comprehensive view of the central phenomenon, climate change, is obtained by complementing messages conveyed through verbal and audiovisual modes. Although we have not discussed the visuals accompanying the research article, we can assume that there are strong intersemiotic relations between the visual and the verbal modes. The visuals used in the article—mainly figures and tables—complement the verbal mode insofar as they contain data that is used as empirical evidence to support the arguments conveyed in the text. We refer the reader to the following case study, where we provide a more detailed commentary of visual verbal interactions within the research article genre (Case study 2).

Understanding the textual voices that represent different viewpoints and perspectives has required a close analysis of the processes of knowledge transformation and strategies of compression, expansion, and persuasion, as we have intended to show in this case study. Through the analysis of one of the various linguistic resources available for the expression of modality and intersubjective stance, the use of modal verbs, we have also identified three competing discourses within the genre network. First, the voice of the institution assuring the commitment of all nations to combat climate change in an assertive tone. Secondly, the voice of the experts showing that actions taken to date may not prevent the consequences—i.e. increasing sea level rise. Finally, the persuasive voice of the blogger presenting an imagined future scenario. This polyphony of voices—the first conveying bare assertions, the second foregrounding likeliness of the predictions expressed through logical extrinsic modal meanings, and the third one accepting the likeliness of hypothetical situations—shapes the lexical and syntactic layers of the texts. Some texts are characterized by a simple and accessible language and others are built upon a lexically dense, syntactically complex discourse style with grammatically compressed phrased structures that package the information. Thus, this network of interconnected genres is characterized by a distinct kind of heteroglossia, or more precisely polyglossia given its polyphony of voices (Bahktin, 1986b; Prior, 2009), encapsulated in three orders of discourse: the institutional, the scientific, and the para-scientific.

From a theoretical standpoint, this case study paves the way for further enquiry into the extent to which the editorial and the short summary act as intermediary genres that facilitate the uptake of the scientists' research article by the rhetor of the blog post. We have labeled both the editorial and the short summary as instances of "scientific discourse" but they actually use a slightly different discourse style from that of the research article, being far from the expository or informational writing type that characterizes the research article and one which is more suitable for wider audiences. For reasons of space, we have not addressed this, but it is an aspect worth investigating. As Tachino (2012: 458) remarks, inquiry into intermediary genres can enhance "our understanding of how multiple genres relate and interact through uptake"—in the context of this book, genre uptake in a digital environment.

Case study 2. Science gatekeeping and add-on genres

In this case study we examine genre interrelations and multisemiotic meaning making in a genre network that addresses issues of disease risk in natural communities caused by increasing global temperatures. Unlike the previous case study, a distinct feature of this network is that it integrates only the perspectives—or voices—of rhetors who are insiders belonging to the scientific community: the authors of an original research article and the journal gatekeepers, that is, the editors and peer reviewers. The genre network has been taken from *eLife* and comprises both expert-to-expert genres and a lay genre. The selection of this case study is also motivated by the implications it poses for pedagogical praxis, especially for the integrated learning of academic literacy, digital and multimodal composing skills.

As shown in Figure 4.2.1, two of the genres in this network, an original research article (6,487 words) and its associated abstract (150 words),[6] involve, as in Case study 1, high-stakes science produced by scientists and disseminated in a high-impact-factor journal online. Both are exemplars of genres for internal communication of science, that is genres that target the expert community. Hyperlinked to them are procedural genres of science gatekeeping: the editor's decision letter, including the reviewers' comments (1,084 words), and the authors' response to them (2,152 words). These genres represent open occluded genres related to the process of research evaluation (Swales, 1996, 2004). In addition, the research article is hyperlinked to two emerging digital genres for article enhancement. The first is the article's impact statement, or statement of significance.[7] Impact statements consist of a single sentence of approximately 250 characters that states the significance or novelty of the article and its value for both the scientific community and society. The impact statement for the selected article totals 23 words. The second emerging digital genre is the digest, which consists of a plain language summary or lay summary of the research article contents. The digest exemplar analyzed in this case study amounts to 249 words. The *eLife* journal platform indicates that the impact statement is written by the article authors and the digest is written by the article authors in collaboration with the series editors.

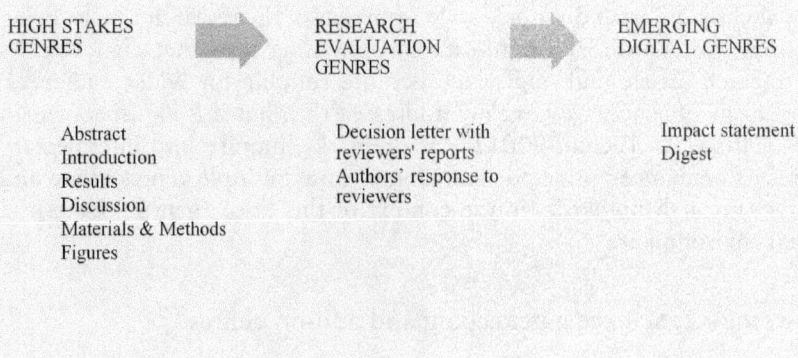

Figure 4.2.1 A network of science gatekeeping and add-on genres.

First, we comment on the salient textual aspects of the genre exemplars contained in this network. We are particularly interested in the analysis of genre interrelations traced through intertextuality—including intertextuality across visual/verbal texts—and interdiscursivity features, following Bhatia (2004), Swales (1990), Swales and Feak (1995), and Jones and Hafner (2012). For the analysis of descriptive statistics relating to this network we draw on corpus analytical procedures, specifically frequency lists, collocates, concordance analysis, and other measures of lexical density and grammatical and syntactic complexity, retrieved with Wordsmith Tools (Scott, 2008). We also retrieved the phraseological profiles of the texts with KfNgram (Fletcher, 2002–2007). The textual material analyzed comprised 10,322 words and the four figures included in the research article.

Text-internal features

If we follow the chronological order of text production, the first text to be described is the original research article. This genre exemplar instantiates the clear-cut regularities of the article genre in terms of rhetorical organization, namely an introduction, the reporting of results, a discussion and materials and methods, or what is conceptualized as the Introduction—Methods—Results—Discussion (IMRaD) pattern for articles in experimental fields (Swales, 1990, 2004). The left-hand-side menu available in the journal platform allows navigation across these sections, the impact statement, and the digest, while the top menu gives access to the decision letter and the authors' response letter.

The article introduction follows the Create-a-Research-Space (CARS) model typical of experimental article introductions and its recurrent moves/steps at the meso level of discourse organization (Swales and Feak, 2009). In Move 1, *Establishing a territory*, the authors set the research scenario and point out the relevance of the topic, in this case, the spread of infectious disease resulting from "unprecedented change as the climate is warming" (Step 1). Afterwards, the authors claim the relevance of the study in the socio-historical context ("a central issue in the 21st century and beyond") (Step 2), and finally give arguments for the relevance of the "disease ecology" framework (Step 3). In Move 2, *Establishing a niche*, the authors explicitly state a gap in the existing knowledge ("Yet, we have a poor understanding of how this framework operates under natural conditions ..."). This is followed by Move 3, *Occupying the niche*, proposing a research process to address the limitations previously expressed ("Here, we test the relative contributions of these three mechanisms through which environmental conditions can drive infectious disease risk"). The remaining rhetorical sections of the article adhere to the expected conventions at the meso level (move/step level) of information organization, as described in the literature (Swales, 2004).

The research article shows some especially frequent collocational patterns. These patterns indicate that the text relies heavily on complex noun phrases formed by noun heads with noun premodifiers (e.g. "host community structure," "host community pace-of-life") and prepositional phrase fragment postmodifiers such as "the relationship between" and "the effect of"). These are all "nominal/phrasal features" (Biber and Gray, 2016: 317) that point to the kind of "discourse densification" (Leech et al., 2009) observed in academic prose and that help the authors fulfill the informational demands of this genre. Such demands are also reflected in the indices of lexical variation and syntactic complexity (TTR = 19.5537; overall mean in words per sentence = 21.27 respectively). As shown in the examples (our emphasis highlighted), the immediate co-text of these highly frequent phrasal patterns further confirms that the text relies on the use of style conventions typical of academic writing, such as authors' self-references, research process verbs ("explore," "test," and "estimate"), markers of epistemic modality ("can"), and different types of dependent clauses such as *to*-infinitive clauses, *that*-complement clauses, means/end clauses, and hypothetical clauses.

> We assessed *the relationship between* abiotic conditions and community pace-of-life by fitting linear mixed models […]

> These results indicate that warming temperatures can modify *the effect of* host community pace-of-life on disease risk, which we attribute to a change in *the relationship between* host traits and host competence across environmental conditions.

The language used in the editor's decision letter and the authors' response is formal and similar to the formal tone of the research article. Measures of syntactic complexity in these two texts are similar to those of the research article (an average mean (in words) per sentence of 22.430), but this is not the case of measures of lexical diversity (TTR = 4.55, STTR std. dev. = 52.94), which can be explained by the difference of the situational contexts and the purposes of the genres. Notwithstanding this, these texts and their pre-text show semantic interrelations, as shown in the especially frequent content words that they share, such as "elevation," "temperature," "effects," "disease," and "community." However, whereas the article mainly relies on complex noun phrases with nominal pre-modification, these mediational texts are mainly built upon *of*-prepositional phrases (e.g. "in figure #," "between elevation and," "on disease risk"), and noun phrases with prepositional phrase fragments (e.g. "the legend of," "the relationship between"), signaling grammatical elaboration rather than grammatical compression. The situational context accounts for such grammatical choices, appropriate for an overt "conversation" between the authors and the editor and reviewers. This is also made explicit through the especially frequent use of the first-person pronoun "we," accounting for a total of 47 occurrences, 46 of them corresponding to the authors' response. The collocate analysis shows that "we" combines with action and thought verbs in R1 position, such as "replace," "revise," "discuss," "clarify," or "incorporate." The deictic "this," an L1 collocate, encapsulates the meaning of the prior utterance, creates discourse coherence, and builds the argumentation, as in this example:

> To address this, we revised the text throughout the manuscript to focus more explicitly on [...]. To do this, we replaced elevation with temperature and soil moisture in our LMMs.

The second-person pronoun "you" occurs a total of 31 times. Normalized frequencies of this pronoun in the decision letter and the authors' response score 14.77 and 6.96 per 1,000 words respectively. Through the use of "you" and the emphatic use of the auxiliary "do," the editor addresses the objections raised by the reviewers and what the revision of the manuscript should entail (e.g. "state clearly how you did the analysis," "you did measure temperature," "but you did not add temperature"). The reviewers' overt statements pointing out the "limited implications for climate change" are expressed in a polite tone (e.g. "The accurate interpretations of ... would require ...," "The revision should either tone down the claim on global change or supplement analyses ..."). The authors acknowledge the value of the reviewers' comments by specifying how they have addressed the weaknesses of their work (e.g. "[...] we have completely re-analyzed the data, focusing more explicitly on ...," "We further edited Figure 4 (the SEM) to more clearly show ..."). This is in fact the pragmatics of politeness expected in the scholarly discussion of the review process of a research article.

Genre interactions

To examine multisemiotic meaning making across these hyperlinked genres, we track intertextual links and identify different types of recontextualisation. In this case we find the two types of intertextuality defined in Chapter 3, manifest intertextuality and constitutive, or generic, intertextuality, as well as features of interdiscursivity. Turning first to manifest intertextuality, it is worth noting that the article draws heavily on intertextual references to prior texts. All these references take the form of non-integral or research-prominent citations (Feak and Swales, 2009: 45), which is a convention of research article writing in the experimental sciences. Intertextual references in the form of integral (or author-prominent) citations or quotations do not occur, since these citational types are not conventional patterns in research writing in the sciences. To refer to prior sources, the article authors summarize, paraphrase, or make generalizations on the source texts. These discoursal strategies give prominence to the topic investigated, demonstrate the authors' knowledge/expertise in the topic (e.g. "Infectious disease is strongly influenced by host community structure and abiotic conditions (Halliday et al., 2019, 2020a) ..."), and strengthen the authors' claims regarding the findings of the study (e.g. "Our study reveals strong evidence that increasing temperature associated with lower elevation can directly influence disease risk, which we attribute to well-established effects of abiotic conditions (Avenot et al., 2017; Garrett et al., 2006; …").

Traces of manifest intertextuality reveal further interrelations across the genres forming the network. For example, the impact statement and the digest contain intertextual references to the contents and viewpoint expressed in their pre-text, the research article. The journal gatekeeping genres contain references to texts providing the context of the article, which gives visibility to the open research evaluation process. The article is internally hyperlinked to several para-texts, that is, texts within and around it, such as the authors' bios with their credentials, information on data availability, online comments, mentions, and tweets and retweets.

Intertextual traces further reveal the existing intersemiotic relations within the network. Taking frequency lists as a point of departure, we can find shared content words in the abstract and the impact statement. Broadly speaking, both genre exemplars convey similar semantic meanings to those of the article, as deduced from the explicit use of key nouns such as "host community structure and disease," "elevational or environmental gradients," and "disease risk prediction." Table 4.2.1 shows that the predominant discourse style of the abstract is phrasal rather than clausal, as it relies on complex noun phrase structures (coded with Atlas.ti and marked between brackets in this table), pointing to discourse densification. The impact statement shares three of the top highly frequent content words of the research article, namely "host community" (occurring more than

Table 4.2.1 Intertextual traces across related genres (under Creative Commons)

Abstract	Impact statement
[Quantifying the relative impact of environmental conditions and host community structure on disease]^{NP} is [one of the greatest challenges of the 21st century]^{NP} (…). [Both increasing temperature and shifting host communities toward more fast-paced life-history strategies]^{NP} are predicted to increase disease, yet [their independent and interactive effects on disease in natural communities]^{NP} remain unknown. Here, we address this challenge by surveying foliar disease symptoms in 220, 0.5 m-diameter herbaceous plant communities along a 1,100 m elevational gradient. We find that [increasing temperature associated with lower elevation]^{NP} can increase disease by […] determining which host species are present in a given location, and (3) strengthening [the positive effect of host community pace-of-life on disease.]^{NP} These results provide [the first field evidence, under natural conditions, that environmental gradients can alter how host community structure affects disease].^{NP}	*Environmental* gradients can modify a fundamental relationship between *host community structure* and *disease*, with implications for predicting *disease* risk in a changing world.

100 times), "disease" (more than 90 times), and "environmental" (over 50 times), and the only two content words of the abstract in the frequency list, "disease" and "host" (highlighted in italics in the table). However, unlike in the article, the authors compress the contents within a grammatically elaborate clause that makes it easy to grasp the value of the study reported in the article. To take up the article contents and repurpose them, the impact statement uses an overtly assertive tone. In addition, the text conveys evaluative meanings by highlighting the relevance of the subject of the study ("a fundamental relationship between") and its ensuing implications.

Instantiating the second intertextuality type, generic intertextuality, or interdiscursivity, the digest borrows the rhetorical organization of the abstract. This text first gives readers a context, then specifies the research niche, reports the main results and finally concludes with implications of the findings. Yet, the rhetors here take a different discursive orientation. In the first paragraph they move from the general to the specific in order to contextualize the study and provide a rationale. However, while the digest sounds "scientific," mainly because of the use of specialized terminology and, from a grammatical standpoint, because of the fact that the discourse is mainly built upon complex noun phrases with noun premodification (e.g. *"climate change," "changing weather"* patterns, *"disease-causing pathogens"*), it also exhibits plain language. The syntax is simple, and the discourse has a high degree of explicitness, as seen for example in the presence of encapsulating pronouns ("this"), transition and exemplification markers ("For one thing ... for another ...," "for example"), reformulations, and clarifications ("This means that ...," "and so ...," "This is because ..."). This suggests that in the process of recontextualizing specialized contents, the scientific terminology used in complex noun phrases (NPs) has been replaced by more general content words such as "disease," "risk," "pathogens," "traits," "infectious," "plants," "temperatures") (with a frequency of use of at least 5 ocurrences) and refocused to shift the emphasis towards "host communities" to "disease risk," the two most highly frequent content words in the digest, as shown in the following extract.

> Climate change is causing shifts in the ecology and biodiversity of different world regions at unprecedented rates. Global warming is also linked with changes in the risk for certain infectious diseases in humans [...]. There are several possible mechanisms for this. For one thing, changing weather patterns may affect how pathogens grow and reproduce. For another, the distribution ranges of animal and plant hosts of certain disease-causing pathogens are changing because of global warming. This means that the distributions of pathogens are also changing, and so is the severity of the diseases that they cause.
>
> Increasing temperatures may also influence the physiological traits that make host species suitable for pathogens. This is because the traits that

allow species to survive or adapt to changes in their environment may also make them better at hosting and transmitting the pathogens that cause disease. For example, in plant communities, [...].

In contrast, in subsequent paragraphs we find discourse features that are frequently used in scientific prose when authors aim to metaphorically create a research niche ("Despite a lot of research into the effect of climate, it remains unclear ...") and fill the niche by stating the purpose and results of the authors' study ("To investigate this ...," "The aim was to ..."). The highly frequent phraseological patterns retrieved from this text with KfNgram—i.e. "showing that the" and "results indicate that"—instantiate generic intertextuality or interdiscursive appropriation of this discourse type.

> Despite a lot of research into the effects of climate, it remains unclear how temperature, pathogen growth and reproduction, and host species' traits and distributions combine and interact to alter infectious disease risk, especially in wild plant communities. To investigate this, Halliday, Jalo and Laine studied an area in southeast Switzerland where natural temperature and biodiversity change gradually through the region. The aim was to explore how relationships [...] and to understand whether environmental or biological factors influence infectious disease risk more.
>
> Halliday, Jalo and Laine measured the levels of fungal diseases found in the leaves of plant communities spanning 1,100 meters of elevation, showing that higher temperatures increase disease risk both directly and indirectly. [...]. The results also indicated that temperature can affect how the traits of plants drive the transmission rates of fungal pathogens [...].

Furthermore, a third type of discourse can be traced in the closing paragraph of the digest, which contains linguistic features that recur in the discourse of advertising. In the digest the rhetors refocus the contents of their article to overtly highlight the significance of the study and its impact on the prevention of infectious diseases resulting from climate change. This is textualized in a less lexically dense promotional discourse than that of the Discussion section of the article, as shown in the extract.

> This study represents the first analysis, in wild plants, of how changing temperatures, the traits of shifting host species, and resident parasite populations interact to impact infectious disease risk. The insights Halliday, Jalo and Laine provided could aid in predicting how global climate change will influence infectious disease risk.

Visual/verbal interactions

In this case study, we observe that the use of two semiotic modes, the verbal and the visual, reflects the authors' efforts to highlight the study findings

Table 4.2.2 Intersemiotic relations (i)

Extract from article	Associated visual and legend

Disease ecology provides a framework for achieving this goal through careful examination of interactions among hosts, parasites, and the environment (Johnson et al., 2015a; McNew, 1960; Seabloom et al., 2015; Figure 1a). Yet, we have a poor understanding of how this framework operates under natural conditions, in part because several mechanisms can operate [...]
We hypothesized that three non-mutually exclusive mechanisms would determine how environmental conditions influence disease risk in host communities: (1) directly, by altering parasite growth and reproduction (i.e. through abiotic constraints; Figure 1b, path a), (2) indirectly, by altering which host species occur in which locations (i.e. mediated by shifting host community structure; Figure 1b, paths b and c), and (3) indirectly, by altering how host traits influence parasite transmission (i.e. moderated by altering the relationship between host traits and host competence, <u>which we refer to as the trait-competence relationship</u>; Figure 1b, path d).

(a)

Community competence:
Host species vary in their contribution to parasite transmission

Host

Ecological filtering:
Shifting abiotic conditions alter the composition of host communities

Parasite

Environment

Abiotic constraints:
Shifting abiotic conditions alter parasite replication and growth

Trait-competence relationship:
Abiotic conditions modify how host species contribute to parasite transmission

(b)

a. Abiotic constraints

Environmental gradient

b. Ecological filtering

Host Community Structure

c. Community Competence

Disease

d. Trait-Competence relationship

Figure 1 Relationships among hosts parasites and their environment at the scale of host communities. (A) The **disease triangle** (McNew, 1960) **suggests** that a combination of host, parasite, and environmental factors will influence whether disease is observed in a given location. Here **we conceptualize the disease triangle** at the community level as consisting of three overlapping or interacting factors to demonstrate how the influence of environmental gradients on disease risk might depend on how these factors overlap. **We highlight three potential processes that** might occur in these areas of overlap **but acknowledge that** other processes likely occur in these areas as well. (B) **Conceptual metamodel** of an environmental gradient directly influencing disease risk (path a), and indirectly influencing disease risk, both by altering host community structure (i.e. mediation; paths b and c), and by modifying how host community structure influences disease risk (i.e. moderation of the relationship between host traits and host competence, <u>which we refer to as the trait-competence relationship</u>; path d).

and contextualize the research process. Both modes are intertextually linked and share the core content words of the article—i.e. "disease," "host community," and "temperature." The first visual in the article and its associated legend perform an explicitation function, that is, both are used to further conceptualize and clarify the authors' initial hypothesis (the potential processes that might occur) and propose a metaconceptualization (Table 4.2.2).

The claim "Disease ecology provides ..." is summarized in the visual with which, as stated in its legend (see highlighting in bold), the authors "conceptualize the disease triangle," and offer a possible interpretation ("The disease triangle (*McNew, 1960*) suggests that ..."). Figure 1b, where the article text is recontextualized in a "conceptual metamodel," extends the information of the article text and graphically displays the interactions within this model. The legend takes up and paraphrases this extended information in the following way. First, the hypothetical statement of the article text, "We hypothesized that three non-mutually exclusive mechanisms would determine ..." and the asides between brackets including clarification of contents ("i.e. ...") are transformed into an overtly assertive personal statement introduced by the first-person plural pronoun ("We highlight three potential processes that ...") and a hedged coordinate sentence ("but we acknowledge that other processes likely occur ..."). The visual and verbal messages exhibit different clines of authorial positioning towards the utterances expressed in the texts. The boxes/labels in the diagrams display keywords in all the texts—"host community," "environment," "disease," and "risk"—and rephrase the contents of the corresponding extract of the article text. They do so using the same syntactic patterns as those used in the article text (simple Subject-Verb-Object (SVO) or Subject-Verb-Adverb (SVA) patterns) and visualizing relationships with arrows.

The second visual (Figure 2 in Table 4.2.3) also complements the written text in the article and extends it by representing topological relations and giving accurate information about the exact physical locations ("meadows") where the study was conducted, and their arrangement within the site. If we compare the written extract and the visual, there is a sharp contrast between the lexical density and syntactic complexity of the 92-word length sentence and the simplicity of the visual that recreates synthetically the physical environment of the research.

The third visual (Figure 3 in Table 4.2.4) elaborates on the corresponding written extract by providing a synthetic view of its contents, strictly focusing on the salient study findings. The visual mode reproduces the two variables in the horizontal and vertical axes, conveying the full set of results from the factor analysis. In contrast, the extract contains a first-person plural pronoun "we" followed by a research process verb ("performed") and a statement of finality expressed by a non-finite infinitive clause ("to assign ..."). This is followed by a detailed account of the salient findings and abundant references to statistical data, which are explanatory of the different percentages of variance regarding host pace-of-life. In the subsequent text, the relationship between the two variables is commented on in detail and framed within the existing literature by means of non-integral citations. Concurrently, the legend further expands on the different parameters of the analysis and their relationship with "host pace-of-life and disease."

Table 4.2.3 Intersemiotic relations (ii)

Beginning of Results section	Associated visual and legend
To evaluate abiotic constraints on parasite replication and growth (i.e. direct effects), shifting host community structure (i.e. mediation; *Baron and Kenny, 1986*), and modification of the trait-competence relationship (i.e. moderation; *Baron and Kenny, 1986*) as mechanisms through which environmental gradients can influence disease risk, we surveyed 220, 0.5 m-diameter vegetation communities (i.e. small plots), that were established in four meadows along a 1,101 m elevational gradient as part of the Calanda Biodiversity Observatory (CBO) in 2019 in order to investigate biotic and abiotic drivers of species interactions (*Figure 2*; *Supplementary file 1a*).	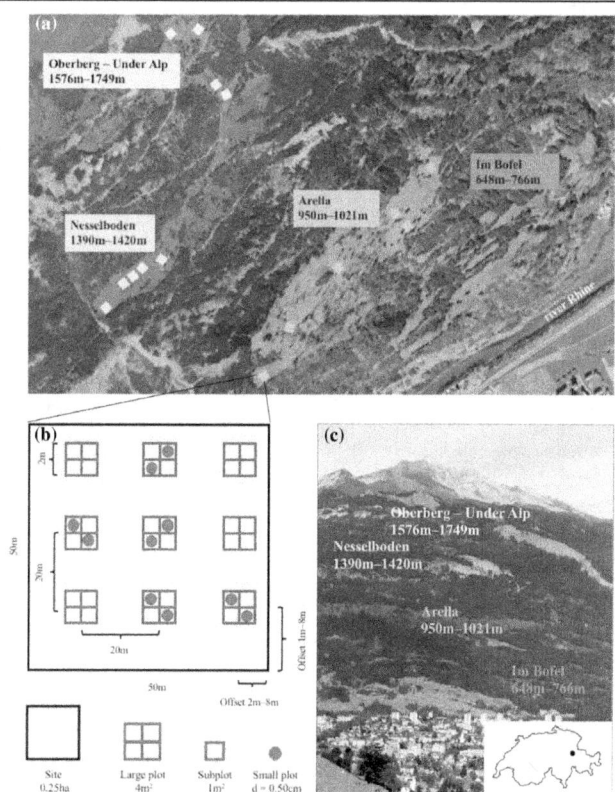

Figure 2 Overview of the Calanda Biodiverstity Observatory. (A) Study meadows and sites on Mount Calanda. Photo: Federal Office of Topography SwissTopo 2020, editing: Mikko Jalo (B) Example of the arrangement of large and small plots within a site. (C) The study meadows on Mount Calanda.

Like the third visual, not illustrated here for space constraints reasons, the fourth and last visual in the article summarizes the main study findings graphically, but the legend is also used explicitly to tell readers how to interpret the semiotic meaning expressed in it ("Colors are drawn to highlight the statistical interaction between host community pace-of-life and temperature") and to give information on the research process, specifically how coefficients were estimated. The objective account of methodological aspects, mostly reported in the passive voice, complements the focused and

Table 4.2.4 Intersemiotic relations (iii)

Extract from Results section	Associated visual and legend

We performed confirmatory factor analysis to assign six foliar functional traits associated with the worldwide leaf economics spectrum to a single axis representing host pace-of-life. [...] This resulted in a single factor, explaining 62% of the variance in specific leaf area, 51% of the variance in leaf chlorophyll content, 25% of the variance in leaf nitrogen, 10% of the variance in leaf phosphorus, and 2% of the variance in leaf lifespan (χ^2 (df= 5) = 4.24, p = 0.52; CFI = 1.019; *Figure 3—figure supplement 1*) [...] Although host community pace-of-life was unrelated to soil moisture (p = 0.13), host community pace-of-life declined with reduced soil-surface temperature associated with higher elevation (p = 0.010; Marginal R2 = 0.11; Conditional R2 = 0.83; *Supplementary file 1b; Figure 3—figure supplement 2*), consistent with expectations regarding shifting host community structure (*Descombes et al., 2017; Hulshof et al., 2013;* but see *Pellissier et al., 2018*). These effects were qualitatively similar when we included soil temperature or air temperature in place of soil-surface temperature in the model, though the effect became marginally nonsignificant when we replaced temperature with elevation in the model (p = 0.066; *Supplementary file 1b; Figure 3—figure supplement 2*).

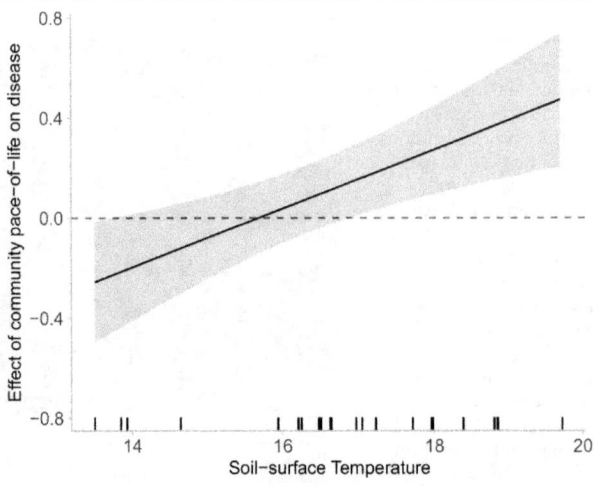

Figure 3 Effect of host community pace-of-life on disease as a function of increasing soil-surface temperature. Model estimated effects of soil-surface temperature on the slope of the relationship between host community pace-of-life and (square-root-transformed) parasite community load (i.e. the interactive effect of host community pace-of-life and soil-surface temperature on disease, which represents a changing trait-competence relationship), estimated from the raw (i.e. unstandardized) coefficients of the linear mixed model testing effects of environmental conditions, community structure, and their interaction on disease. The rug along the x-axis shows the distribution of the empirical data. Communities that experience the highest soil-surface temperatures (i.e. located at the lowest elevation) exhibit the strongest positive relationship between host pace-of-life and disease. That positive relationship weakens as temperature declines, and below mean-soil-surface temperatures of 17.5°C (i.e. above 1,000 m), there is no relationship between host pace-of-life and disease.

elaborate commentary of the findings and their significance provided in the corresponding fragment of the text that makes references to the figure. The extract contains markers of self-representation ("we," "our"), abundant references to detailed statistical data (that replicate exactly those of the figure), as well as paraphrases explicitly signaled by the discourse marker "i.e." We do not illustrate this last visual-verbal interrelation further for reasons of space but we refer the reader to the article to verify this interrelation.

Interpretation

In this genre network we observe several ways in which, echoing Jamieson (1975), genres establish connections with other genres and creatively mix different genres. We offered evidence in this case study of genre connections as traced through features of intertextuality and interdiscursivity. On the one hand, the comparison across the genres allows us to understand that, in addition to sharing semantic meanings, interdependent genres within the network entail instances of manifest and constitutive or generic intertextuality. Instances of manifest intertextuality mainly occur in the research article and, in terms of the cross-genre comparison, at the level of discourse semantics, as evinced by the range of keywords shared by the article, its associated abstract, the editor's letter and authors' response, and the add-on genres of this enhanced publication, namely the impact statement and the digest. Further, the genre connections within this network confirm that the process of genre remediation proves to have no impact on the form and substance of these two prototypical genres, as argued earlier (see Harmon, 2019). Both their overall rhetorical macro structure for information organization, their organization at the meso level (i.e. rhetorical moves and steps) and the linguistic resources used in both genres adhere to the standard conventions. The generic integrity of the two research-evaluation genres is the product of the conventional features that the members of the scientific community recognize as prototypical of these genres.

This case study has further shown that, in keeping their generic integrity, hyperlinked genres differ among themselves in several ways. As seen above, although all these texts address the topic of climate effects, the approach to the contents expressed in the article (and associated abstract) and the research evaluation processual genres is different from that of the two emerging digital genres, being most noticeable at the level of discourse pragmatics. The textual construction of the article and its abstract and the situational context of these genres is constrained by the hierarchical relationships established in the scientific community, influencing language choice. Language variation thus confirms that "the schematic structure of the discourse" (Swales, 1990: 53) does pose constraints at the level of vocabulary, grammar, and syntax. As seen earlier, in this case study the situational context and the socio-historical context both constrain the article writers' discourse style—mostly factual, or evidence based, with features of pragmatic politeness signaled by the use of hedges and boosting resources—as well as the discourse of the genres involved in the research evaluation process—an assertive, unhedged discourse, but expressed in a collegial and respectful tone.

On the other hand, in illustrating the application of the genre lens, this case study also offers a better understanding of issues of genre innovation in web environments. For example, through the hyperlinking affordances of the digital medium the article contents are enhanced by means of two emerging digital genres: the impact statement and the digest. Furthermore, whereas the formal regularities observed in the journal article reflect stabilized conventions, from the findings of the present analysis it appears that genre innovation in emerging genres connected online with traditional genres mainly relies on appropriation of both generic features and discourses other than scientific discourse. As seen in this case study, the impact statement acts as a mini genre hyperlinked to the article to promote the article contents, and it does so exhibiting the same semantic profile as that of the article. As another example, the digest repurposes the semantic sharedness with the article contents but transforms it by borrowing some of the discourse features of popularization genres with the purpose of reaching broader publics. The rationale behind the features of interdiscursivity and the blend or merging of discourses that results from processes of knowledge re-entextualization in these two emerging genres lies, precisely, in their distinct communicative goals and target audiences (specialist *vs* non-specialist). This case study provides evidence of strong complementarity between the article and its add-on, constellating genres for article enhancement purposes and content dissemination to broader publics.

This case study also addresses intersemiotic relations within research articles online. The visual/verbal interrelations identified in this study are mainly relationships of extension, as defined by Van Leeuwen (2008), since in the research article one mode (the visual mode here) adds related information to the other mode (the verbal mode). In this specific case, both modes complement one another, which creates strong relationships within a single genre, the research article, as also deduced from the semantic meanings shared by the legends of the visuals and the article text. Making the two modes complementary makes meaning making in the high-stakes genre more effective. Furthermore, it is worth noting that because the visual messages are part of a high-stakes genre, they are neither polysemous nor do they appeal to emotions, as is the case, for instance, in case studies 4 and 5. In short, the visual messages are consistent with the expectations of the genre: to make readers fully and accurately informed about the research processes and outcomes.

Moving to a broader canvas, this case study illustrates that exploring a genre network demonstrates that Web 2.0 communication makes it necessary to analyze individual genres working in conjunction with other genres to better understand multisemiotic meaning making in online environments. The descriptions of individual genres in the network and the comparative analysis of these genres at different textual levels informs the analyst about the way the two key criteria for genre identification—intended audience and communicative purpose—determine the form and substance of connected genres. As we argued in previous chapters, the concept of audience becomes problematic when applied to Internet communication because the latter involves a duality of audiences or dual audiences. However, as the case study illustrates, it seems feasible to determine the different situational contexts that a genre network embraces. As seen above, here there appear to be at least clearly distinguishable contexts and "virtual dialogues." One context is that shared by the authors and the members of the scientific community which is enacted by the article, the abstract, and the impact statement. Another context is that shared by the authors and the science gatekeepers, enacted by the two research evaluation process genres analyzed in this case. The third context, enacted by the digest, is one shared by the article authors/editors and broader publics. As observed in the case study, through discoursal uptake and recontextualization, the rhetors repurpose expert scientific knowledge in the form of a digest to accommodate science to broader publics (Fahnestock, 1986).

Case study 3. Science reproducibility genres

In this case study we analyze the intersemiotic relationships between two genres used for reporting methods and protocols in the *Journal of Visualized Experiments* (*JoVE*). We specifically explore these relations by describing the process of recontextualization of the contents of a written article (totaling 5,059 words) in a video article (1,018 transcribed words). The focus of this case study is therefore the examination of resemiotization. Specifically, we compare a video methods article (VMA hereafter) that describes a protocol for producing particles to study coronaviruses with the associated written methods article (WMA hereafter) on which the video is based (Millet et al., 2019).[8] *JoVE*[9] is an online journal that publishes scientific research in video format, thus enabling authors "to dynamically present their methods, data analyses and results clearly, accurately, and professionally" (*JoVE* homepage, 2015). The video is produced by *JoVE* scriptwriters and videographers from the manuscript submitted by the authors, whose final version is also published on the *JoVE* website in the form of an original written article.

The WMA and the VMA are embedded in the hypermodal platform of *JoVE*, where there are two hyperlinked web pages (the WMA web page and the VMA web page), each of them in turn incorporating various genres. Figure 4.3.1 shows the genres forming this network.

As seen in Figure 4.3.2, the largest part of the VMA web page is occupied by the video itself, which is displayed at the top of the page. The video is followed by the title of the article and the names of the authors, with hyperlinks to a page for each of the authors where further information about them can be found, including a biography, links to publications, and links to their other video articles in *JoVE*. Below the names, there is a timeline ("Chapters") that allows users to click on specific sections of the video and access them directly, a summary (a shorter and less technical text than the abstract) and the video transcription. On the right bar of the page, there is a link ("Article") to the WMA web page, links to social networking tools like Facebook or Twitter,

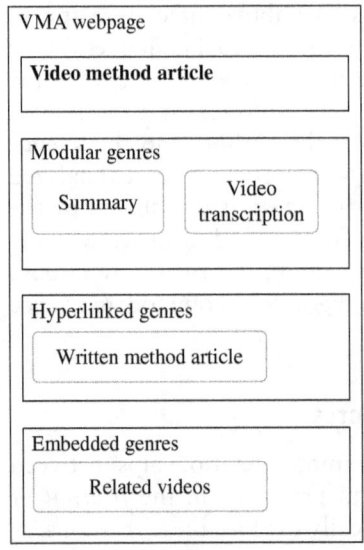
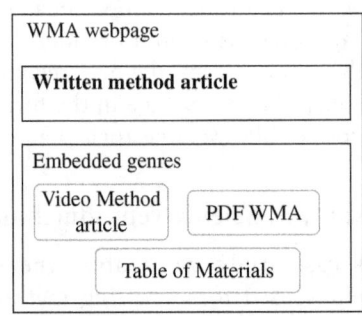

Figure 4.3.1 Ecology of genres in the selected VMA.

Figure 4.3.2 Layout of the VMA web page.

and quick links to "Related Videos." Clicking on the "Article" link on the right leads to the WMA web page. This web page displays the following sections: Summary, Abstract, Introduction, Protocol, Representative Results, Discussion, Disclosures, Acknowledgements, Materials, and References. The longest and most detailed section is the Protocol, which is consistent with the fact that *JoVE* articles are different from conventional scientific articles, since the focus is on the methods and the purpose is to help other researchers to replicate the research. The WMA web page also embeds the VMA and other downloadable documents, such as the article PDF and the Table of Materials. The different genres in the *JoVE* platform are therefore related not only intertextually but also physically, by being embedded on the same website and hyperlinked. On each of the two web pages the user is provided with easy access both to the VMA and the WMA, as well as to other genres that he/she can choose to access specific or further information.

In this case study we combine genre analysis (Bhatia, 1993, 2004; Swales, 1990, 2004) with multimodal discourse analysis (Jewitt et al., 2016; Bezemer and Kress, 2008). We also rely on general descriptive statistics retrieved with Wordsmith Tools (Scott, 2008) to explore the register and discourse features of the two genre exemplars. We first identify and compare the rhetorical structures of the WMA and VMA to determine whether the affordances and constraints of the VMA have an influence on the moves. We then focus on the relations between semiotic resources in the VMA and explore how different semiotic resources are orchestrated to make (experiential, interpersonal, and textual) meaning in this genre. Following Bezemer and Kress (2008) and Van Leeuwen (2008), we also analyze the process of recontextualization involved in moving the contents expressed in the written article to the video article.

Rhetorical organization

Table 4.3.1 compares the rhetorical structures of the WMA and the VMA. As stated previously, the VMA is based on the written article and therefore the main sections and many of the moves in the WMA are replicated. However, the information in the WMA is recontextualized and resemiotized in the VMA through processes of selection, arrangement, foregrounding, and social reposition (Bezemer and Kress, 2008).

The VMA is a much shorter text than the WMA to achieve the pedagogical purpose of this video genre. A comparison of the number of words in each section of both articles reveals a significant difference in the proportion of words devoted to each part in each text (Table 4.3.2). In the VMA the Protocol/Demonstration section accounts for 74.5% of the words, which suggests that the other sections have a secondary function in this text. By contrast, in the WMA the Introduction (and associated abstract), Results, and Conclusion account for 23%, 21%, and 22% of the words, respectively, which points to the complementarity of both texts.

Table 4.3.1 Rhetorical moves in the WMA and the VMA

Written Methods Article (WMA)	Video Methods Article (VMA)
Title	
Abstract	
Purpose of the protocol	
Summarizing the procedure	
Explaining significance	
Limitations	
Introduction	**Introduction**
Providing background information	*Purpose of the protocol*
Purpose of the protocol	<u>Introducing self (main researcher)</u>
Explaining significance	*Explaining significance*
Summarizing the procedure	*Potential applications*
Potential applications	<u>Critical steps in the protocol</u>
	<u>Introducing additional researchers</u>
Protocol/ Demonstration	**Protocol/ Demonstration**
Introducing materials	*Describing steps*
Describing steps	*Drawing attention to important*
Drawing attention to important	*information*
information	
Results	**Results**
Explaining how results were achieved	*Presenting and evaluating results*
Presenting and evaluating results	
Discussion/ Conclusion	**Discussion/ Conclusion**
Summary of the results	*Critical steps in the protocol*
Critical steps in the protocol	*Giving tips: Possible solution to*
Limitations	*limitations*
Giving tips: Possible solution to	<u>Important safety information</u>
limitations	
Potential applications of the method	

underlined moves appear only in the VMA
moves in italics appear in both genres

Table 4.3.2 Distribution of words in each section in the
WMA and the VMA

	WMA		VMA	
	words	%	*words*	%
(Abstract) + Introduction	1199	23	95	10
Protocol/ Demonstration	1685	33	707	**74.5**
Results	1062	21	90	9.5
Discussion/ Conclusion	1113	22	57	6

Like the WMA, the VMA begins with a title frame which includes the title of the article, the authors of the WMA and their affiliation. The introduction, which lasts 40 seconds (90 words), includes only some of the moves that appear in the WMA in the abstract and in the written article introduction (*Purpose of the protocol, Significance of the protocol, Potential applications*), by this means recontextualizing the information of these sections of the WMA in a more succinct way. The move *Providing background information* in the WMA gives basic information to understand the protocols, such as the definition of the type of particle that is produced through the protocol ("Viral pseudotyped particles, also named pseudovirions, are powerful tools that enable us to easily study the function of viral fusion proteins"). However, this move is not included in the VMA, which indicates that both texts complement one another. The move *Purpose of the protocol* is shared by both articles, but the information provided in the VMA is much more concise, less technical, and focuses explicitly on the key aspect of the paper: safety. The WMA provides minute details of why the particles obtained with this protocol can be used safely; the VMA does not provide this information but draws attention to "safety" when presenting the purpose of the protocol (Table 4.3.3).

The introduction in the VMA also includes moves that do not occur in the introduction in the WMA. One of these moves is *Critical steps in the protocol* (i.e. "For best results it is important to monitor cell health and density for transfection as well as for pseudotype virus infection"). While in the WMA this move appears in the conclusion rather than in the introduction, in the VMA this is an important move, with critical information being provided both in the introduction and the conclusion. Other moves not present in the WMA introduction involve the presentation of the researchers (i.e. *Introducing self, Introducing additional researchers*).

Table 4.3.3 Move Purpose of the protocol in the WMA and in the VMA

WMA	VMA
Introduction: The protocol's main purpose is to show how to obtain coronavirus spike pseudotyped particles that are based on a murine leukemia virus (MLV) core and contain a luciferase reporter gene. This feature allows them to be used in **intermediate biosafety level** facilities (BSL-2) and is an important advantage over using highly pathogenic native viruses that require **higher biosafety** facilities (BSL-3, BSL-4 which are not as readily available) when conducting virus entry studies.	This protocol allows researchers to generate pseudotype viruses that can be **safely** used to study viral entry events of highly pathogenic viruses, such as sal coronavirus and most coronaviruses.

The Protocol/Demonstration, the longest and most important section in the video (6:73 minutes; 707 words), consists of a description of the different steps to replicate the method. Verbally it is presented by a voice-over, and the researchers are only represented as performing the action processes listed by the voice-over ("incubate," "add," "aspirate," "move," "mix"), but not verbal processes. In the WMA the protocol consists of a list of steps, many of which are followed by "NOTES" which highlight important information for carrying out the step successfully (i.e. the move *Drawing attention to important information*), as seen in the example.

> Wash cells with 10 mL of pre-warmed (37°C) Dulbecco's Phosphate Buffered Saline (DPBS) twice.
> **NOTE:** Handle HEK293T/17 cells with care as they easily detach.

In contrast with the Protocol Demonstration, in the VMA, the Results and Discussion/Conclusion are presented very briefly (97 and 40 seconds, respectively). In both articles the purpose of the Results section is to provide evidence of the effectiveness of the method. While in the WMA the Discussion/Conclusion consists of several moves, in the VMA the few seconds of the Conclusion are used to emphasize the most important information. Although both the WMA and the VMA include the move *Critical steps in the protocol*, the information provided in this move in both genres is different. The VMA presents two general pieces of advice, not present in the WMA, which are, however, critical to the success of the protocol ("Double check calculations prior to performing the step. And make sure all solutions are mixed well during the step"). The Conclusion in the VMA also includes a move not present in the WMA (*Important safety information*).

Semiosis and recontextualization

Different semiotic modes are combined in multimodal ensembles in the VMA to recontextualize the meanings conveyed in the WMA and achieve the discoursal functionality of the moves. While in the WMA only written text and static images (in the form of tables) are used, the VMA resemiotizes part of the information in the WMA by combining speech, moving images, static images, and written text.

The Introduction is presented by the main researcher, filmed in a close-up shot, and superimposed text is used to display his name and institution. While in the WMA the authors are only represented by their name after the title, in the VMA they are represented linguistically and visually as identifiable social actors. In the first moves of the Introduction the verbal mode makes the largest contribution to meaning. Visually, in all the moves the researcher is presented as a "talking head" in a close-up shot, looking uninterruptedly at the camera with a neutral facial expression and speaking in a monotonous tone, which means that gestures or intonation do not contribute to meaning creation. Speech is the dominant mode, and the viewer focuses only on the verbal message. The last move of the intro-duction (*Introducing additional researchers:* "Demonstrating this procedure will be Tiffany Tang, and Lakshmi Nathan. Graduate students from my laboratory") corresponds to a new close-up shot, with the image of the dem-onstrators. In this shot the demonstrators are working in the lab, wearing lab coats, but when they are mentioned by the main researcher, they look at the audience and smile. The direct gaze and smile have an interpersonal function: they reduce the level of formality and invite the viewers to engage with the ensuing description of the protocol. The way this move is real-ized is quite revealing if we take into account the likely audience of VMAs. According to Pritsker, *JoVE* CEO/co-founder, the audience(s) for *JoVE* arti-cles are "practising researchers, practical scientists who work in the labora-tory. So everywhere between professors to postdoc to graduate students to undergraduate, including technicians" (Hafner, 2018: 24). The appearance of the researchers, working in the laboratory and smiling, reduces the dis-tance between the researchers/authors of the paper and the audience(s). It also repositions some of these authors as demonstrators who seek to help the viewers perform the protocol.

The Protocol section in the VMA consists of the demonstration of the different steps described verbally in the WMA. Thanks to the affordances of the medium, recontextualization in the VMA involves adding visual infor-mation to the linguistic information provided in the WMA. In this section a greater number of modes than in the other sections of the VMA are com-bined to make meaning: the spoken language of the voice-over explaining the procedure is combined with moving images—the researchers demon-strating what the voice-over says—and written text. The different sections of the Protocol are indicated in the same way in both the WMA and the VMA, through frames displaying the title of each of the main procedures within the protocol on a black background (e.g. "Three-plasmid Co-transfection"). However, sequentiality is conveyed differently in the two articles. In the WMA it is conveyed through numbers (i.e. a numbered list of steps). In the VMA sequentiality is also conveyed verbally, but with words/phrases expressing sequence (e.g. "to begin," "then," "next") ("To begin, carefully wash HEK293T cells with 10 milliliters of 37 degrees Celsius, pre-warmed DPBS twice").

The language and visual elements are orchestrated to help the viewer understand how the protocol is carried out. The oral text describes mate-rial processes—e.g. "add," "incubate," "wash," "move"—which are shown by means of moving images. The participants have the role of actors, who demonstrate how the protocol is conducted. The way modes are combined helps to focus the viewers' attention on what the demonstrators are doing. First, the demonstrators are depersonalized: they do not report what they are doing, and they do not look at the camera. The viewer cannot see their facial expressions. Most of the shots are close-up shots but instead

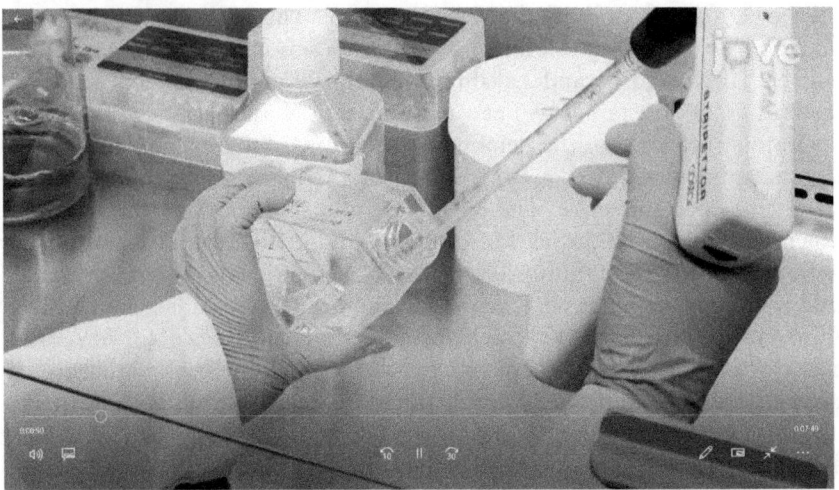

Figure 4.3.3 Screenshot of the Protocol section.

of close-ups of researchers' faces, the viewer sees the demonstrators' hands and the devices that they are working with in the foreground (Figure 4.3.3). The face and gaze of the demonstrators seem to be distractor elements and are thus not essential in the video. The narrative is refocused to give greater prominence to research procedures. The demonstrators' hands and the elements they are working with are made prominent not only by placing them in the foreground, but also by using color contrast: the demonstrators wear blue gloves and some of the components they use are red-colored, which contrasts with the white/gray color of most of the tools in the biosafety cabinet where the protocol is performed.

As Hafner (2018) explains, the visual resources available in the VMA enhance and extend the meaning made in the WMA. Circumstances of the process such as time, frequency, quantity, or temperature are expressed verbally (e.g. "… attach cells with one milliliter of 25% trypsin solution that has been pre-warmed at 37 degrees Celsius. Then, incubate the flask of cells at 37 degrees Celsius in 5% carbon dioxide environment for three to five minutes until cells start detaching"). Among the highest-ranked (i.e. the top 20 most frequent) content words are "degrees," "microliters," and "minutes." The moving images are used to realize other circumstances, such as manner, which are better conveyed through this mode. For instance, in the first step ("To begin, carefully wash HEK293T cells with 10 milliliters of 37 degrees Celsius, pre-warmed DPBS twice"), the spoken text is suitable to express quantities and degrees but does not explain what instrument should be used and exactly "how" these instruments should be handled. This information is provided visually (Figure 4.3.3). Similarly, when the voice-over narrator introduces the following research step formulated by an action verb in the imperative form ("Then, incubate the flask of cells at 37 degrees Celsius in 5% carbon dioxide environment for three to five minutes until cells start detaching"), the video illustrates how the demonstrator places the flask in a machine, thus "showing" where the incubation takes place. Again, the face of the demonstrator is not shown, and the focus is put on the action that she is performing. The subsequent instruction, "[p]lace the tube in a luminometer device. Close the lid and measure the luminescence value of the tube", is accompanied by a frozen image exclusively showing the device and how exactly to measure this value. Images, therefore, enhance the meaning conveyed in the verbal text by providing information on the "how" of the action.

Meaning making also draws on the visual mode to show what the researcher should see when performing a step or the results of a step, by this means recreating the outcome of the research action conducted by the researcher. For instance, while the voice-over states, "and then count cells using a cell counting slide and a light microscope," an image of what can be seen with the light microscope is shown on screen. Similarly, "[T]he color of the medium should be light pink or slightly orange" is accompanied by a close shot of wells with a medium of that color. Thus, the images facilitate access to knowledge not presented in the WMA by adding further details.

Another mode used in the protocol section is written text superimposed on the moving images. This written text is used to make information prominent by foregrounding it. For instance, when the voice-over says "pre-warmed DPBS" the superimposed text "Dulbecco's phosphate buffered saline" appearing in the foreground on the video clarifies the technical acronym. Superimposed text is most frequently used to draw the readers' attention to information that is particularly important for the protocol: "Add transfection reagent to the reduced serum-medium, not the other way around," "Plate can be handled outside the biosafety cabinet," "Avoid displacing the liquid on wall of the tube." As noted above, in the WMA the move *Drawing attention to important information* appears after the description of some steps. This information is made prominent through the word NOTE in capital letters and highlighted in bold type (e.g. "**NOTE:** Perform this step in the biosafety cabinet," "**NOTE:** Ideally, cell density should be in the 40–60% confluency range"). Interestingly, the information foregrounded in the VMA through superimposed text does not appear in the **NOTES** in the WMA, but in the description of the steps.

Superimposed written text is also used to create intertextual links with the written article and to encourage the reader to get information from other genres in the *JoVE* platform. When the voice-over says: "Then add three microliters per well of the lipid-based transfection reagent to 47 microliters per well of the reduced serum cell culture medium," the textual fragment

"See manuscript for media composition" appears on screen. Also, at the beginning of one of the stages, the voice-over informs "to perform three plasmid co-transfection, first mix calculated volumes ...," while the following text on screen explicitly mentions the associated article: "See manuscript for more details," instantiating manifest intertextuality. These written texts, asking the reader to find information in the WMA, show the interdependence between the WMA and VMA. The WMA is intended as the first step to produce the VMA, but the VMA is not a stand-alone text which could be understood without the WMA.

The Representative Results section in the WMA is relatively long and highly technical, where most of the highest-ranking content words are technical words and acronyms ("pseudotyped," "particles," "cells," "infectivity," "COV," "infection," "MERS-SPP," "SARS-SPP"), rather than academic words (see Coxhead, 2000). The text is accompanied by two tables (including numerical information about quantities of reagents) and four figures presenting the results. One of the high frequency words in this section of the WMA is "figure" (15 occurrences; 1.36% of the words in the section), which shows that reporting of results relies heavily on visual elements. The Results section in the VMA is four sentences long, focusing only on the presentation of the main results with the help of the figures. In this section the VMA seems to have adopted discourse conventions of the conference presentation genre. Instead of moving images of the demonstrators, this section displays the figures that appear in the WMA, framed in a slide layout, and combined with a voice-over. While in the WMA the figures are static, in the VMA they are dynamic and change as the voice-over presents the results. As in the WMA, the figures are used to provide evidence that supports the claims presented in the spoken text. All the technical details present in the results sections in the WMA—i.e. numerical information presented in the tables, or the detailed explanation of how the results have been achieved or how the effectivity of the protocol has been measured—have been excluded when moving information to the VMA, which again shows the complementarity between the WMA and the VMA (Figure 4.3.4).

Finally, in the short Conclusion section in the VMA (four sentences) the two demonstrators appear on screen, this time without the lab coat, and looking directly at the camera. While the first time they appear in the video, at the end of the introduction, they smile at the audience, conveying proximity and collegiality, in the conclusion their facial expression is serious, which conveys authoritativeness and helps to create a professional identity. The conclusion section consists of three shots, each one corresponding to one of the moves of the conclusion (Table 4.3.1). In the last shot (where safety information is presented) there is a change of demonstrator. This change and the fact that this is the last information provided foregrounds this information ("While the pseudotyped particles described here are safer surrogates than nato viruses, they still require biosafety level two precautions").

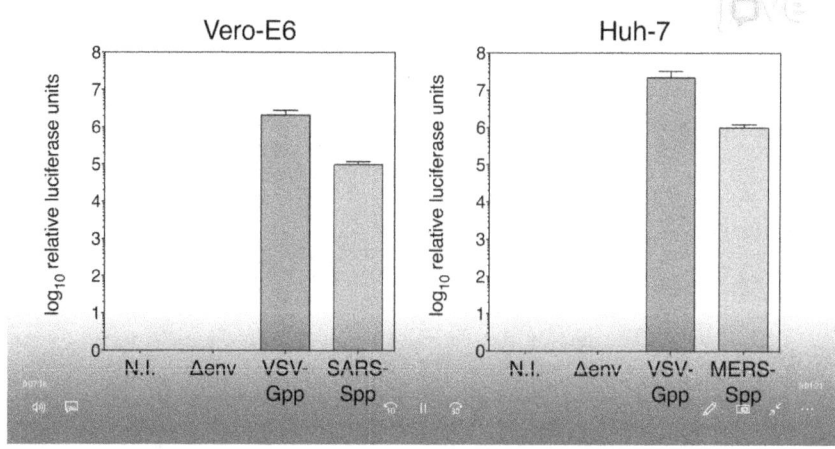

Figure 4.3.4 Screenshot of the Results section.

Features of discourse and register

Register features are similar in the WMA and the VMA. Both are texts with a high frequency of nouns and prepositions, as deduced from the word frequency list. This indicates that both texts have an informational focus (Conrad and Biber, 2001). In this case study, simple corpus searches of features of involved or interactional discourse show that none of the texts display these features, e.g. first- and second-person pronouns, *wh*-questions, emphatics, amplifiers, expression of personal feelings, and attitude (Conrad and Biber, 2001: 24). This can be explained by the fact that, even if the text in the video is spoken discourse, it has been carefully scripted.

Although both texts share features of informational discourse, there are differences in their keyness. As shown in Table 4.3.4, the most frequent content words in the WMA are technical nouns and adjectives that modify them ("cells," "pseudotyped"). In the VMA, however, they are nouns expressing measurements ("minutes," "degrees"), research process verbs in imperative form ("incubate," "add") and, as expected in a process description, adverbs expressing sequence ("next"). Therefore, while in the VMA the most important part is the protocol where instructions are given through imperative forms, in the WMA other types of information are also very important, which suggests text complementarity. Table 4.3.4 further shows that one of the highest-ranked content words in the WMA is "note" (21 occurrences), used to convey the importance of specific information. The word "important" also occurs frequently in the text (17 occurrences), usually in the anticipatory *it* + evaluative adjective pattern followed by a *to*-complement clause (e.g. "It is important to + verbs such as bear/keep in mind/remember/

Table 4.3.4 Highest-ranked content words in the WMA and the VMA (top 20 words)

WMA			VMA		
Rank	*No. of occurrences*	*Relative frequency*	*Rank*	*No. of occurrences*	*Relative frequency*
cells	82	1.52	cells	25	2.57
cell	63	1.17	Celsius	14	1.44
pseudotyped	63	1.17	degrees	14	1.44
particles	61	1.13	well	12	1.23
transfection	37	0.68	incubate	10	1.03
viral	37	0.68	add	9	0.92
well	36	0.67	cell	8	0.82
infectivity	31	0.58	microliters	8	0.82
luciferase	30	0.57	SPP	8	0.82
envelope	27	0.50	then	8	0.82
virus	25	0.46	next	7	0.72
mers	24	0.44	particles	7	0.72
used	24	0.44	carbon	6	0.62
particle	22	0.41	dioxide	6	0.62
assay	21	0.39	light	6	0.62
entry	21	0.39	minutes	6	0.62
note	21	0.39	pseudotyped	6	0.62
MLV	20	0.37	tube	6	0.62
plate	20	0.37	solution	6	0.62
coronavirus	19	0.35	transfection	6	0.62

consider"). The evaluative adjective "key," which is used repeatedly, always foregrounds a specific research step ("This is a key step"). The second most frequent imperative form in the WMA after "note" is "avoid" (12 occurrences), which introduces a purpose statement ("include extra wells to avoid running out of transfection reagent mix"). The VMA, however, does not include any purpose clauses explaining the reasons for an action, and therefore the viewer needs to refer to the WMA to find this type of information.

Table 4.3.4 also shows that the WMA includes more technical words than the VMA. The comparison thus indicates that one of the processes in the recontextualization involves the exclusion of highly technical information. As shown in the example, the first step in the protocol in the WMA is a very detailed description of the materials that cannot be found in the VMA, since it is not possible to convey it effectively using the modes available in the video.

By standard cell culture techniques, obtain an 80–90% confluent 75 cm^2 flask of HEK-293T/17 cells passaged in complete Dulbecco's Modified Eagle's Medium (DMEM-C) containing 10% (vol/vol) fetal bovine serum (FBS), 10 mM 4-(2-hydroxyethyl)-1piperazineethanesulf onic acid (HEPES), 100 IU/mL penicillin, and 100 μg/mL streptomycin.

Interpretation

The *JoVE* website should be understood as a response to the demands of the OS movement for transparency and accessibility which can improve the efficiency and reliability of scientific research (Bartling and Friesike, 2014; Sarcina, 2019). In particular, the genres of the *JoVE* website have emerged to address the "replication crisis," or the fact that it is very difficult, if not impossible, to replicate the experimental results of a large number of studies (Baker, 2016; Munafò et al., 2017). The *JoVE* project is an example of new practices in science communication which emphasize the importance of transparency to facilitate reproducibility and improve the quality of methodology. These practices have led to the emergence of genres that focus on the evaluation of methods—which is also the case of other emerging genres such as the registered report, which reports methods to be reviewed prior to data collection (Mehlenbacher, 2019a)—and on improved reporting of protocols—see the *JoVE* article, or research protocols at protocols .io., an open access repository used by researchers to provide and link to step-by-step reports of the protocols in their published articles (Munafò et al., 2017). In addition, the *JoVE* website also has a pedagogical purpose, since visualization of how a protocol or technique is carried out serves as instruction for researchers who want to learn this new method or technique (Vardell, 2015). The *JoVE* editors themselves explain the rationale of the genre on the *JoVE* website:

> *JoVE* serves the research community as a scientific methods journal for efficient dissemination, reproduction and discussion of experimental approaches in biological, medical, chemical and physical research. Video is an effective publication format as it ensures more efficient transfer of information than traditional text articles. *JoVE* publishes novel methods, innovative applications of existing techniques and gold standard protocols in a scholarly video and text format. Detailed text and representative results accompany every video. https://www.jove .com/authors/peer-review

It is precisely in relation to these new communicative demands that the genres explored in this case study should be framed. The interaction of the genres in the platform, in particular the WMA, the VMA, and the Table of Materials, fulfills one common goal: to give instructions to enable the reproducibility of procedures and replicability of results. Although the WMA and the VMA could be considered two different genres and have their own generic integrity, they complement one another, and the video could not be understood without the WMA and vice versa. As shown in the case study, the VMA has been created by recontextualizing the WMA and taking advantage of the affordances of the new medium to facilitate reproducibility and instruction. Recontextualizing its pre-text in the VMA involved (i) the selection of key information and information that could be best conveyed with visual resources, (ii) visual addition, i.e. addition of visual information to enhance the information given in the WMA, (iii) foregrounding of particularly important information, (iv) rearrangement of information to make some of it particularly prominent, and (v) social repositioning of some authors of the articles as demonstrators seeking to help researchers who may be interested in replicating the protocol.

The process of recontextualization did not, however, involve a change in register or discourse style, which can be explained by the fact that both texts share an informational focus. Other studies of VMAs (Hafner, 2018; Maier and Engberg, 2018) have nonetheless found some examples of language features of involved discourse, especially in the introduction ("my name is," "let's get started"), which suggests that this genre may display colloquial features especially frequent in conversation. This is not the case of the genre exemplar analyzed here. Hafner (2018) points out that in the introduction of VMAs the researchers seek to engage the audience, and they may do this using multiple semiotic resources (e.g. language, researchers' gaze, embodied action). Here, the researchers' gazes and smiles in the introduction also served to reduce formality and engage the audience, which supports Hafner's (2018: 32) claim that one of the innovative features of VMAs, when compared with the WMAs on which they are based, is the construction of engagement in the introduction, which is "likely linked to the visual affordances of the video medium." Therefore, even if the discourse of both the WMA and the VMA is informational, recontextualizing the WMA into the VMA seems to involve adding and constructing engagement in the introduction.

The analysis of the rhetorical organization, discourse resources, and register in this case study reveals the importance of selection and foregrounding when producing the VMA: only the most important moves present in the WMA, conveying key information, are kept. Information is more concise and less technical, and key aspects such as safety issues are emphasized. The longest part of the VMA is that for which the affordances of the visual mode are most useful to achieve the purposes of the genre: the protocol, where images help to communicate protocol subtleties that are difficult or impossible to convey in the written text. The multimodal affordances of the medium are harnessed in the VMA to combine speech, moving images, static images, and written text, and thus help to convey transparency and facilitate the reproducibility of research. This case study offers evidence of the way the VMA modes are combined to draw the viewers' attention towards what the demonstrators are doing, with the visual mode enhancing and extending the meaning conveyed in the verbal text. In the protocol the language and visual elements complement one another, thus helping the viewer understand how the protocol was carried out. Speech follows linear/sequential logic (Jones and Hafner, 2012), and is therefore well suited for reproducing the sequence of instructions provided in the written article.

The moving images complement the text by providing information on the "how" of the action. The images also facilitate access to knowledge not presented in the WMA by adding explanatory details (e.g. color), since, as Lemke (1998: 87) notes, image is suited to representing gradation (of texture, color) and spatial relations between objects. The superimposed written text also contributes to meaning making by directing the viewer to the WMA or foregrounding important information.

This case study also illustrates that the affordances of digital media enable a variety of simultaneous relations between genres. The WMA and the VMA are interdependent and could be considered as components of a genre encompassing both of them: the online article. The VMA takes up the information given in the WMA and resemiotizes it to add a visual component that facilitates reproduction of research. Both are intertextually linked, since the VMA is constructed through the selection of information from the WMA, but also materially connected: the two genres are embedded in two highly interconnected (through several hyperlinks) web pages or digital layers in the journal platform. The reader may use the hyperlinks to move easily between the genres in the article to related web pages, and to move within and beyond the platform to other texts with further information. The video provides the visual demonstration of the protocol, but it is embedded in a web context together with other genres that provide the information necessary to understand the video (i.e. the WMA), or to increase related knowledge—e.g. links to the authors' information, which in turn link to their publications, and related videos. Hypertext and genre embedding enable readers/viewers to choose their own trajectory to access the information in the various genres accessible through the platform.

Case study 4. Participatory science genres

In this case study we explore processes of knowledge recontextualization in a Citizen Science (CS) project. The selected project PollinatorWatch, addresses the topic of climate effects on pollinators.[10] It is available on Zooniverse, the largest CS web portal. We examine the internal intersemiotic relations of the CS genre as it contains embedded genres that are internally hyperlinked. We also describe processes of meaning making and repurposing within this single digital genre. It will be seen that a CS project can be conceived of as a macro genre made up of various interrelated micro genres (mini texts) and embedded genres. The principal aim of this case study is to illustrate how both the language used in internal genres of science communication (e.g. research articles) and the language used in external genres of science communication (e.g. popularizations) merge in the CS project genre.

Figures 4.4.1 and 4.4.2 illustrate the project homepage, showing the various texts inserted in the modular layout of the web design as well as the different genres embedded in some pages and subpages.

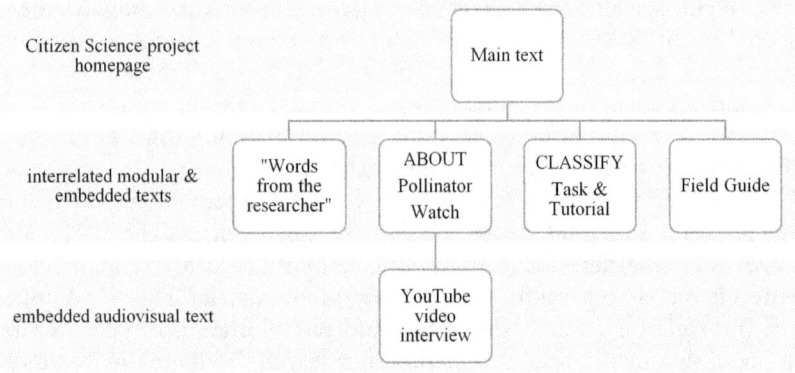

Figure 4.4.1 Website structure of the CS project.

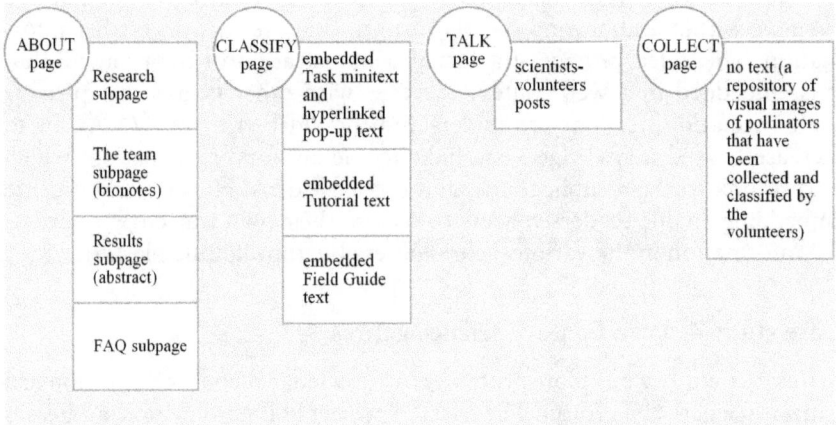

Figure 4.4.2 Main menu pages and subpages of the project website.

In this case study we combine the perspectives of genre studies (Bhatia, 1993, 2004; Swales, 1990, 2004) and register studies (Biber et al., 1999; Biber and Conrad, 2020; Biber and Gray, 2016). We analyze the textual material included in the project website, accounting for 2,968 words. We use corpus-based methods for the automatic extraction of frequency lists and concordance lines with Wordsmith Tools (Scott, 2008). We also complemented this methodological approach with inductive coding of grammar and syntax features at a phrase/clause level, which amount to a total of 37 grammatical/syntactic codes and 1,581 quotations. Code-document co-occurrence indicating relations between code grounded value and the number of quotes in a given text, and code co-occurrence, or occurrence of two codes in the same textual fragment, were identified with Atlas.ti v8 to track inter-semiotic relations in this genre.

To address visual/verbal interrelations, we apply the perspective of multimodality (Jewitt et al., 2016) to analyze the 35 static images included on the website (31 photographs of flowers and pollinators, four photographs of the researchers, and a flow diagram).

Informational layout

The information given in this project does not follow the logic of a linear text. Rather, it is organized in modular texts, which indicates that one important constraint in digital communication is the fixity of the homepage design, imposed by the materiality of the medium. The homepage displays just a two-sentence text with a relevant image in the background with which the researchers invite the interested publics to collaborate in the project (Figure 4.4.3). A low-quality picture of a crop of white flowers, one of them with a pollinator, is used as a background image and the favicon is a white flower with a yellow pollinator. Right below the written message, there are two buttons (Learn more and Get started) internally hyperlinked to two tabs of the homepage navigation menu, About and Classify.

Scrolling down, we find two modular texts. One is located in the central pane. This is a three-paragraph text entitled About PollinatorWatch. It contains an embedded genre, a YouTube video called "Specialist interview," in which the head of the research team explains the research procedures conducted to identify pollinators. To the left of this central text there is a two-sentence text entitled "Words from the researcher" highlighting the global scope of the project and its intended outcome.

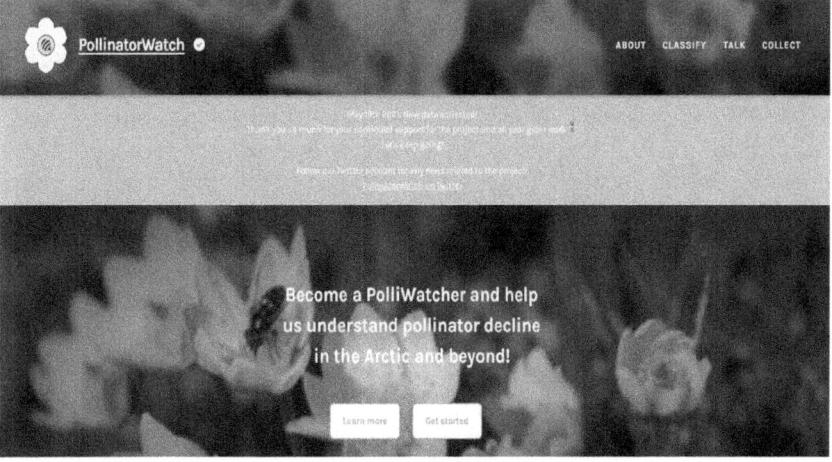

Figure 4.4.3 PollinatorWatch homepage and navigation menu.

The navigation Menu contains four pages: About, Classify, Talk, and Collect. The About page splits the information into four subpages—Research, The team, Results, and FAQ. These texts have different rhetorical functions. In the Research subpage the text is organized as follows. In the first paragraph, the researchers inform about the scope of the project and its main goals. In the following paragraph, they provide details of the current and forthcoming phases of the project. This description is accompanied by a visual that synthesizes the text—a flow diagram summarizing the three project stages and the research processes involved in each of them. The description of each stage appears along with the image of the favicon. Shades of yellow are used to better identify each phase visually. The researchers then inform about other related projects that have stemmed from the group's research, which gives it credibility. The last paragraph is an exhortative message. The researchers thank the volunteers by acknowledging their collaboration ("You are helping us study Arctic pollinators in a new way and for that, we are very grateful. Thank you!").

The Team subpage includes four bionotes describing the researchers' credentials. They are written in the third-person singular, and all refer to academic merits and research awards. The bionotes are located right below the photographs of the four researchers with smiling faces. These are close-ups where gaze, camera angle, and distance convey close personal contact, following Kress and Van Leeuwen (2006) and Jones and Hafner (2012). The researchers look directly ahead at viewers, aiming to build trust.

In the Results subpage, the reader finds the abstract of a published article—a Perspective article authored by the project leader—that also reinforces the researchers' credibility. Finally, the FAQ subpage consists of a typical list of questions to which the researchers respond by giving detailed and concise explanations and clarifications. This is an open text, in the sense that it can be updated during the duration of the project, thus evincing the openness of this digital genre.

The Classify page embeds two interrelated mini texts, Task and Tutorial. The first text uses a direct question ("Do you see any pollinators in the image?") to invite action on the part of volunteers, who need to click on the Yes/No buttons and Done/Done and Talk buttons to complete each classification task. Below the Yes/No buttons we find the message "Need some help with this task?" that is hyperlinked to the Field Guide, where volunteers find several pop-up mini texts that resemble dictionary entries. Each entry includes short definitions of different pollinators, false positives, and flowering stages ("budding," "blooming," and "seeding"). Each entry also includes images of specimens that visually clarify the key concepts the volunteers need to know to classify the images of the Task. The text refers to these images as "example images" (Figure 4.4.4).

The Classify page also contains a third embedded genre exemplar, the Tutorial, which consists of an interactive slide show presentation. Volunteers need to click on the Continue button to navigate the slides or move back and forth. The function of this text is key insofar as it explains

Budding

Budding is the first stage of the flowering process. In this early stage the petals of the flower are not yet visible, and the pollinators will not have access to the pollen inside. We want to count the pollinator, so you should still click the **yes**-button when you see a pollinator visiting a bud. However, if you see a bud with no pollinator, you should click the **no**-button

Here are some example images of a budding flower:

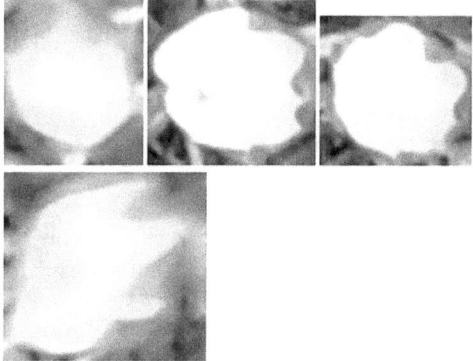

Figure 4.4.4 Example of entry from the Field Guide.

the low-quality real images that appear concurrently with the text and give volunteers succinct instructions on how to identify and classify them. The Tutorial text defines the images as "small pieces of larger time-lapse images that a computer has determined looks like flowers" and that have been zoomed in many times "to spot the small pollinators." The blurred pictures of flowers in some of the Tutorial slides reinforce this message. In all the slides, the image is accompanied by a text that creates a persuasive appeal, encouraging volunteers not to get discouraged and foregrounding once again the important value of the volunteers' participation ("even though you might not see a lot of pollinators, you are still doing valuable work towards uncovering their ecological role!").

Finally, the Talk page has in-built interactive tools that support inter-action between the volunteers and the researchers. The language of these interactions contains colloquial features such as first-person pronouns and features of online interactions such as contracted verb forms and emoticons. Volunteers' comments and researchers' responses resemble a chat or a blog post, making this page a virtual space where the researchers and volunteers co-construct knowledge informally by discussing doubts in classification, queries, and technical problems found in the classification process.

Language variation across texts

Starting with the short message on the project homepage, the researchers first address volunteers by extending a polite, informal invitation built upon a very simple syntactic structure—"Become a PolliWatcher and help us understand pollinator decline in the Arctic and beyond!" (see Figure 4.4.3). The message is constructed upon two coordinate directives, "a speech act expecting some action from the addressee" (Biber et al., 1999: 456), an explicit self-reference to the research team through the oblique form "us," and an expression of strong feelings formulated by an exclamative clause. The coordinating conjunct recalls the add-on strategy of conversational language. Each independent clause contains a single idea with two distinct actions, taking a decision ("become") and supporting the researchers' understanding of climate effects on pollinators worldwide ("help us understand").

The About PollinatorWatch text contains features that are also especially frequent in academic prose, such as complex noun phrases and non-finite *to*-infinitive clauses. At the same time, it exhibits features that are especially frequent in conversation, for example simple noun phrases, the second-person pronoun "you," and exclamatives. The merging of expository and conversational discourse styles is illustrated in the following extract.

> For the past three years time-lapse cameras set up at various locations across the Arctic have gathered millions of images of flowers. The goal is to monitor the periodic changes in plant and insect life and ultimately to train a computer to automatically locate and identify pollinators in the images. [...] Now we need your help spotting the pollinators that visit the flowers!

In the Words from the researcher text, the pronoun "we" is used not only to refer to the research team but also to all researchers in this field, emphasizing the global scope of the project. The project outcome is formulated in a very explicit, grammatically elaborate noun phrase in subject position ("An automatic system for identification [...] would be a game-changer for that kind of diversity studies"). The use of parallel coordinating sentences ("We see on a global scale that populations of insects are broadly in decline and we have limited understanding of why that is") also recalls the language of conversation. In the embedded YouTube video, the researcher uses informal language. His speech is characterized by especially frequent simple noun phrases and adverb phrases, self and others' references ("I/we/you" pronouns), *to*-infinitive clauses expressing purpose, and the linking adverbial "so" introducing deductions (Biber et al., 1999: 449), as the extract illustrates. The add-on strategy, (Biber et al., 1999: 228) marking conversational turns in complex utterances, can be taken to reflect "the interactive character of speech" that characterizes the researcher's interview.

> We collect about 2,000 insect samples each season each summer and I think roughly, you can say, it takes about an hour to process one of these samples. and that is to just sort them into broad taxonomic groups. [...] actually, there are very few keys or literature support to actually identify these groups so an automatic system for identification of insects more broadly would be ...

The merging of grammar features frequently found in two very distinct registers—academic prose and conversation suggests interdiscursivity in the processes of discourse uptake and recontextualization, and more specifically in the reorganization and refocusing of specialized knowledge. The Words from the researcher are intertextually linked to both the About PollinatorWatch text and the embedded YouTube video interview, as shown in Table 4.4.1. As deduced from the lexical profile of these texts, each text has a different degree of conceptual depth. A hypernym is used in the Words from the researcher text ("populations of insects") while hyponyms are used in the About PollinatorWatch text to provide examples ("pollinators such as bees and other insects"). Discipline-specific words such as "samples" and "specimens" are only used in the Words from the researcher text and the YouTube video interview. Across these three connected texts we observe examples of reformulation strategies. The Words from the researcher text, shown below, summarize the meanings expressed in the other two texts to make two broad claims concerning the decline of pollinators resulting from global climate change and the knowledge gap, or "limited understanding." Both are used as arguments to further claim the need for "an automatic system for identification" of these insects.

> We see on a global scale that populations of <u>insects are broadly in decline</u> and <u>we have limited understanding</u> of why that is. <u>An automatic system for identification</u> [...] would be a game-changer for that kind of diversity studies.

As shown in Table 4.4.1, in the other two texts the arguments are more elaborate, although the focus differs slightly. The About PollinatorWatch text first places the focus on the research process and situates it geographically ("the Arctic") and then explains the importance of automatizing it

Table 4.4.1 Traces of intertextuality across texts

Extract from About PollinatorWatch	Extract from video interview
For the past three years time-lapse cameras set up at various locations across the Arctic have gathered millions of images of flowers. The goal is to monitor the periodic changes in plant and insect life and *ultimately to train a computer to automatically locate and identify pollinators in the images.* [...] Currently, researchers are working hard to collect and identify insect samples <u>in the hope of gaining a better understanding of how and why these species change in response to global change.</u> *Automating some of this process frees up time for interpretation, consequently* <u>leading to a deeper understanding of the underlying mechanisms.</u> Your participation will <u>help researchers gain a better understanding of the consequences of global change</u> and help them spread this knowledge more widely.	It's **quite a tedious** process. It takes **a long time** in the field, but it also takes **a substantial amount of time** in the lab, and it requires **quite considerable expertise knowledge** about these specific organisms, specific traits that allow us to identify them down to the species level. We collect about 2,000 insect samples each season each summer and I think roughly, you can say, **it takes about an hour to process one** of these samples. And **that is to just sort them** into broad taxonomic groups. The subsequent identification of the insects down to species level can take anything from yet another hour to **maybe even ten or 20 hours** for one sample and, typically, this means that we don't only identify a subset of the specimens that we collect for some groups. <u>Actually, there are very few keys or literature support to actually identify these groups.</u> *So, an automatic system for identification of insects more broadly would be a game-changer for that kind of diversity studies.*

(highlighted in italics). Like the Words from the researcher text, it also explicitly mentions three times the need for a better understanding of global change on pollinators. Here the first-person pronoun is not used. The reference to "researchers" suggests that this is not a knowledge gap in this research team but in the research community as a whole, hence the invitation to collaborate with them in the classification process. In the YouTube video interview, the focus is not the research gap—which is only very tacitly mentioned (as underlined in the extract)—but the very time-consuming, "tedious" research process that the researchers undertake when locating and identifying pollinators. This was also briefly mentioned in the About PollinatorWatch text, but in the video this is expanded in much greater detail, which is actually one of the functions of the discourse of Citizen Science: to transform scientific meaning into meaning accessible for volunteers (Reid, 2019). In this project the researcher provides an evaluative account of the procedures used both in the field and in the laboratory. The first-person pronoun "we" makes the arguments personal. Evaluative statements abound (highlighted in bold). It is only at the very end of the text that, building upon the researcher's compelling reasons, the claim for an automatic system for identification of insects is made.

The text of the Research subpage has a high presence of simple noun phrases. Adverb phrases and non-finite verb phrases encapsulating *to*-infinitive clauses are also especially frequent and are used for building arguments and providing explanations to highlight the importance of the project goal. These non-finite clauses are used to convey stance meanings making overt the researcher's perspective towards the utterances. Many are epistemic, expressing "how true is the information of the clause" (Biber et al., 1999: 356), and they reinforce the idea conveyed in the clause, as in "To *actually* identify these groups ...," or "This in turn allows us to spend more time on the thing that *actually* matters." Some are attitudinal, "expressing the speaker's evaluation and attitudes towards the content of a clause" (Biber et al., 1999: 356), as is "*In fact* the training requires large quantities of data" and "*Hopefully* the new network will return images containing pollinators" (our emphasis added).

In the Team bionotes the researchers refer to themselves in the third person to show their credentials and build a professional identity, but they use their first names or short names and not their family names. A simple syntax is used (S+V+Attribute or S+V+O+A clauses), as in "Toke is the principal researcher on the project," or "Isa is doing her internship on citizen science at the University of Tromsø." This language contrasts with the grammatically compressed discourse style of the abstract embedded in the Results subpage. The abstract is characterized by the use of complex noun phrases containing prepositional phrase post-modifiers or dependent clauses. These are grammar features especially frequent in academic prose. The use of these features in the abstract reflects the informational demands of this genre. These grammar and discourse style features differ from those of the FAQ subpage. In this subpage the researchers respond to queries and give explanations and clarifications to volunteers. The tone is informal and the discourse style, which can be described as pedagogical, reflects the rhetorical constraints underlying any list of frequently asked questions and answers.

The Task, Tutorial, and Field Guide embedded in the Classify page show "colloquial features associated with present day conversation" (Biber et al., 1999: 312), although further differences can be traced between them. The Task is a fixed template that displays the straightforward question "Do you see any pollinators in the image?" Below, volunteers find Yes/No buttons and right below them a very straightforward message expressed by means of an interrogative clause ("Need some help with this task?"). As in informal conversation, both the auxiliary "do" and the subject pronoun "you" are left out (Figure 4.4.5). This message is hyperlinked to a pop-up text that further elaborates and clarifies what the volunteers need to do in the Task, this time expressed in a more explicit way (e.g. "Your task is to determine if

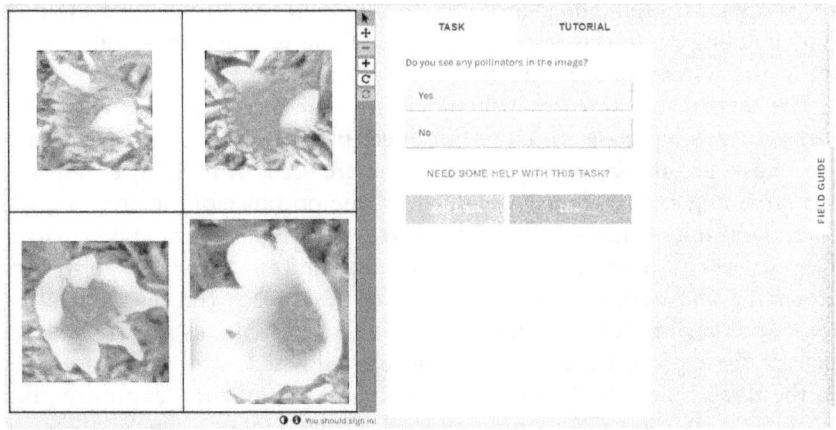

Figure 4.4.5 Screenshot of Task, Tutorial, and Field Guide in the Classify page.

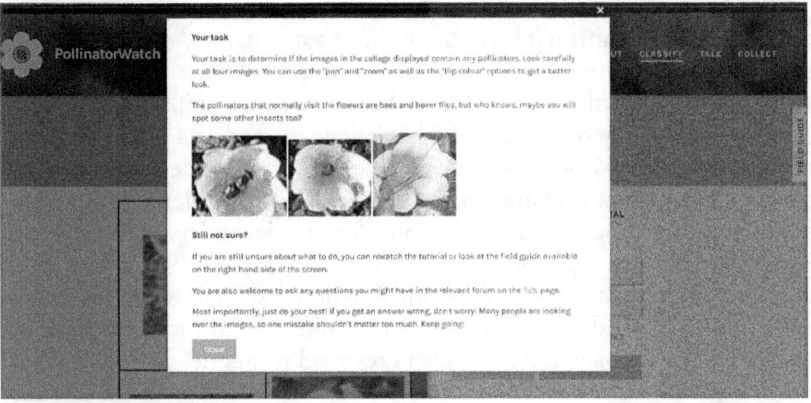

Figure 4.4.6 A pop-up text sample of Task.

the images in the collage displayed contain any pollinators") (Figure 4.4.6). This pop-up text includes a link to the Tutorial ("If you are still unsure about what to do you can re-watch the Tutorial ..."). The main language features of the Tutorial are simple verb phrases, imperatives (directives) and simple noun phrases containing the second person pronoun "you." This text has a clear pedagogical function, as its main function is to provide further guidance to the volunteers.

In all the Field Guide entries, copula *be* verb-phrase structures convey technical definitions—e.g. "Technically speaking a pollinator *is* any animal that transfers pollen grains from flower to flower." Verb-phrases with subordinate reason-result clauses and *that*-complement clauses and relative clauses introducing clarifications are also especially frequent. Adverb phrases function as deictics (e.g. "here," "below") to refer to visuals (photographs), and for clarification ("Here are some example images of a blooming Dryas Integrifolia without pollinators"). There is also greater lexical diversity, as seen in the presence of hyponyms of "animal," such as birds, buds, pollinators, insects. Once again, images play an important didactic function. They provide exemplification to support volunteers in the classification task.

Finally, the language of the Talk page contains features that are especially frequent in online interactions, such as first and second person singular pronouns "I/you" and, syntactically, interrogatives and, above all, exclamatives in the researcher's responses, which express interest, understanding, and enthusiasm. Emoticons also convey emotions and feelings, as illustrated in the volunteer's query:

> I've seen a few insects that are not described in the tags (though I'm sure they're included in etc.) is there any sort of list that I can refer to as reference to better classify? i.e. should I be looking out for wasps or yellow jackets? I hope this makes sense on paper, let me know if I need to clarify.

And in the researcher's response:

> Hi @Parker-Delaney
> The most frequently occurring insects can be found in the field guide under the "Urban Green Roofs" category
> If you see something that isn't in there, classify it as "other insect"
>
> and by all means let us know in "Talk" 😃
> All the best,
> Isa

The use of an informal, conversational style indicates the researcher's interest in creating proximity with their volunteer, especially when the volunteer fails to classify the image. In this case, the researcher encourages the volunteer to keep on with the classification task using exclamative clauses ("Not sure if that is a pollinator on the top left panel!—I think it is just the stamen of an 'old' flower. Keep up the good work!").

Interpretation

While in other case studies we look at sets of connected genres online forming networks, here we have examined a single macro genre, or "complex whole," the term used by Kelly and Maddalena (2016) to conceptualize the citizen science genre, given that it contains internally hyperlinked modular texts/mini texts and embedded genres. Essentially, we have analyzed how the internal organization of this genre exemplar is constrained by the fixed architecture or layout of the web portal. We have seen how the technological affordances of the portal recreate different virtual situational contexts and thus give researchers opportunities for informing, persuading, instructing, and interacting with the citizen volunteers. In light of the findings, we would argue that the rhetorical organization of this macro genre evokes, metaphorically speaking, an implosive mosaic of internally hyperlinked connected texts and complementary multisemiotic modes—the visual, the verbal, and the aural. Using such a textual mosaic is a way of making the messages conveyed in these texts more comprehensible to all participants. Whether the mosaic of texts makes it easier for the volunteers to engage in Citizen Science projects, or whether it is simply the outcome of the fixity of the platform or the materiality of the medium, as Yates and Sumner (1997) call it, requires further exploration in future research.

This case study further shows that some of the genres—mini genres—embedded in this macro genre constitute a colony of interrelated genres. The Task, the Tutorial, and the Field Guide fit into Bhatia's (2004: 59) definition of colony, a grouping of genres "serving broadly similar communicative purposes, but not necessarily all the communicative purposes in cases where they serve more than one." The genres of this colony have a common communicative purpose, which is to support volunteers in the classification process. This explains why they share some rhetorical strategies, discourse features, and visual elements. Yet, as seen earlier, they do this in different ways, through the specific discourse features that are part of their generic integrity. The Task texts are instructional. Directives are used to tell volunteers what to do. The discourse functionality of the Tutorial text is to explain to volunteers how to do the tasks, and the Field Guide entries describe and illustrate the images the volunteers need to classify.

Another genre colony is formed by the modular texts on the project's website. The homepage message invites volunteers to collaborate on the project. The "About this project" text succinctly contextualizes the project, and this information is complemented by the Words from the researcher text and the embedded YouTube video interview. The shared communicative purpose of this genre colony is to provide readers with a rationale for the researchers' interest in getting citizens involved in the project. But while the homepage message is very persuasive, the remaining texts provide matter-of-fact arguments, repurposing and refocusing meanings in various ways to bring to the fore the research process, including the time-consuming process of classifying pollinators. A third genre colony is found in the About subpages, whose texts revolve around the context and goals of the project, and all inform readers but do so in different ways. The Overview text situates the research territory and highlights the importance of the scientific contribution the project will bring with it. This information is complemented by information on the professional profile of the researchers (Team bios) and publication outputs (Results). These texts exhibit different functionalities, as previously noted.

Another theoretical conceptualization illustrated by this case study is the intermediary genre (Freadman, 2002). One could argue that the article abstract inserted in the Results subpage might act as an intermediary genre to re-entextualize and repurpose meanings from the different texts on the project website. The rhetors can be said to have engaged in a process of recontextualization of expert knowledge, transforming it into knowledge accessible to non-specialists. The high lexical density and complexity associated with the compressed discourse style observed in the embedded abstract contrasts with the elaborate discourse style, mostly clausal rather than phrasal, of the remaining texts and with the informal language of the FAQ subpage and Talk page, typical of online interactions (Luzón and Pérez-Llantada, 2022; Sindoni, 2013). Evidence based on the grounding of the inductive coding analysis further suggests that the more technical the language, the less syntactically elaborate the discourse style is. Therefore, the observed reformulation strategies may account for the merging of registers and discourses, which lends credence to the hybrid nature of the CS project as a para-scientific genre.

On the other hand, the information organization and linguistic variation across the modular texts and embedded texts in this macro genre illustrate how the genre constrains the way language can be used within a single genre to fulfill multiple functionalities, as discussed earlier. As Swales (1990: 40) explains, "[r]egisters impose constraints at the linguistic levels of vocabulary and syntax, whereas genre constraints operate at the level of discourse structure" and the genre determines the ways in which register varies within it. Here register appears to impose constraints at the levels of vocabulary and syntax in different ways. In fact, the register variation observed in this macro genre indicates that the researchers, as rhetors, reshape and use/manipulate the genre affordances to refocus and recontextualize contents to accomplish different communicative goals. The merging of a phrasal, grammatically compressed style and, at other times, a highly conversational usage of language indicates that when researchers engage in this genre, they need to address several communicative intentions, i.e. to inform interested citizens about their project, to actively engage them in scientific enquiry, to educate them in issues of science and, more broadly, to build trust in scientific research. In short, the CS project homepage showcases the merging of expository, didactic, and persuasive discourse types which expand meanings to support the citizens/volunteers in the classification process. Therefore, features of interdiscursivity or the merging of discourses in science popularization genres, described previously in the literature (e.g. Motta-Roth, 2009; Motta-Roth and Scherer, 2016), also appear to be distinct features of this macro genre.

Multimodality or the interrelation of visual and verbal modes plays a key role in this macro genre, each mode complementing the other. The images of flowers and pollinators are typological, insofar as they mainly serve to illustrate the types of pollinators the volunteers need to classify. They reinforce the explanations and definitions provided by the verbal mode and are used to disambiguate volunteers' doubts in the classification process. In fact, we have seen that this project website contains many informative and instructional visual elements. The range of visuals—from the homepage visual to the images of the genre colony formed by the Task, Tutorial, and Field Guide genres—serves to flatten out possible knowledge asymmetries between the researchers and their citizen volunteers. Further, we have seen how the multimodal elements are multifunctional insofar as they support and complement the way the researchers express their ideas and use language to achieve different communicative goals. In the case of the photos of the Team bionotes, the case study illustrates how digital media enable researchers to construct social (professional) identities that create proximity towards volunteers. Added to this, it is also worth recalling the functionality of the embedded YouTube video recontextualizing the information of the About PollinatorWatch text but putting the emphasis on the real settings in which the research processes take place. While the researcher appears at the forefront explaining how time-consuming their classification tasks are, in the background we see the team's workspaces and labs, which confer greater credibility and trust to the researcher's explanation. By stressing the laborious and complex process of classifying pollinators, the researcher tacitly claims the need for an automatic classification system.

The layout design of the web portal determines different dominance relations across the visual and verbal modes. We can recall here that the About PollinatorWatch text containing the embedded YouTube video, placed in the central pane of the homepage, represents the dominant element, if we apply Jones and Hafner's (2012) multimodality framework. Notwithstanding, while Words from the researcher might then be considered a marginal element, the fact that it is strategically placed on the left-hand side pane gives the text a degree of dominance. The intertextual link between the two texts, namely the reference to the need for understanding why populations of insects are broadly in decline and the background image of the blurred flowers with pollinators, suggests that the type of image/text interaction is one of concurrence. The messages conveyed by the two modes are the same and reinforce each other.

More broadly, in taking a text-first approach, in this case study we have briefly addressed the social and historical aspects of the rhetoric inherent in this genre (Jamieson and Campbell, 1982; Miller, 1984). The action accomplished by the genre is fulfilled by interdiscursively intertwining genres, registers, and modes. This can be a rhetorical response to the rationale underpinning the genre, namely, to make science accessible to non-specialist audiences, to democratize science, to increase scientific literacy, and to fulfill research goals ("to develop an automatic system for identification of pollinators based on deep learning and computer vision to transform the entomology field," as the abstract of the Results subpage states). It is not difficult to empirically validate the claim that the genre has accomplished its social action. As the project statistics available on the project homepage state, on 100% completion, the project has involved 7,071 volunteers, 542,668 classifications, and 167,733 subjects classified.

To better understand the generic nature of CS projects, we therefore need to bear in mind text-external factors (Bhatia, 2004), namely social structures (professional practices and constraints), actors, and participant relations. The rhetorical and linguistic features of the project described previously can be accounted for by the specific situational and rhetorical context of the genre, a text type that responds to asymmetries between participants (in terms of knowledge background) and the social exigences of Citizen Science, representing an approach to science communication that encourages citizens' participation in the research process (Fecher and Friesike, 2014). As seen in this case study, to get citizens involved in such a process, the researchers draw upon the technological affordances and design of the web portal and draw upon language resources to tell their volunteers how to participate and why their help in identifying and classifying data is important. Self- and others' references, taking the form of *I, you,* and *we* pronouns, are encapsulated in simple noun phrases. Verb phrases with dependent clauses and adverbial phrases conveying stance construct an overall grammatically elaborate rather than grammatically compressed discourse style that performs several communicative functions. Firstly, these language choices convey the researchers' perspective on the topic and its global impact. Secondly, they serve to elaborate and reformulate specialist knowledge in different ways to make it clear to volunteers the scope of the research and what the classification tasks involve. By this means, the genre's social action is fulfilled.

In the spirit of Open Science, the genre's social exigence is to build trust and interest in science and support the democratization of science by reaching broader and diversified publics, increasing their scientific literacy (Bonney et al., 2009). This is mainly done through a mosaic of texts, as noted earlier, namely, the Overview text, the About PollinatorWatch text, the YouTube video, and the About page texts. The social exigences of promoting participatory science may also explain the inclusion of mini genres such as tasks (with which the volunteers can actually apply scientific knowledge), the tutorial (with which they can learn in more detail how to engage in the research process), and the field guide (with which they can increase their knowledge about types of pollinators and flowering processes), all of them with potential for increasing citizens' scientific literacy. Furthermore, in reinterpreting, reorganizing, and refocusing contents to highlight different facets of the project and its outcomes and to engage citizens in scientific research, the researchers' underlying ideologies and values regarding the impact of climate change have come to the surface, at times more overtly, at other times less explicitly, laying bare their positioning as specialists towards climate effects on pollinators as a global concern. From the way the project is constructed, we can infer that researchers' communication practices for participatory science align with the democratization of science and its educational scope, and also satisfy their own professional and personal agendas, namely using volunteers' data classification behavior to train computers using deep learning methods.

Case study 5. Para-scientific and social media genres

This case study focuses on the multimodal and semiotic recontextualization processes in news articles and tweets that comment on an original research article. It aims to illustrate how contents taken from a highly standardized genre, the research article, are repurposed and recontextualized to make them suitable for the new context and address the information needs of a diversified audience. We analyze a research article on lake stratification—entitled "Phenological shifts in lake stratification under climate change" (Woolway et al., 2021),[11] totaling 5,338 words and published on 19 April 2021 in the journal *Nature Communications*—along with other non-academic genre texts that explicitly mention this article. The information on the article metrics provided by the journal, based on Altmetric data, reveals that by 19 June 2021 this article had been mentioned in 106 tweets, one blog, and three news articles.

The blogpost that mentions the original article—Skeptical Science New Research for Week #16, 2021—is a news aggregation post which offers a hyperlinked list of articles on climate change published in a week, among them the article by Woolway et al. (2021), to bring them to the attention of the readers. Of the three news articles that mention the original article, the first is an example of what Bray (2019) calls "backgrounding texts," i.e. texts that use the article as evidence or background information for other issues, hence entailing processes of repurposing. The text, entitled "Renewable Energy Is Suddenly Startlingly Cheap" and published in *The New Yorker* by the American environmentalist and journalist Bill McKibben, includes a paragraph on the *Nature Communications* (NC hereafter) article, reporting one of its main claims and including a quotation dealing with economic effects. The second news article, published in *Carbon Brief*[12] (CB hereafter) is what Bray (2019) calls a "foregrounding text," i.e. a text that foregrounds the original article and its findings to achieve certain rhetorical effects. The text is written by Ayesha Tandon, a science journalist specializing in climate science. The third news article, published on the World Economic Forum website (https://www.weforum .org/), reproduces the *CB* text, including the note "This article was first published in *Carbon Brief*." The author provides a link to the article in *CB*, which shows how copy and paste is a practice that helps dissemination in online environments. To examine how the original research article is taken up by different rhetors, we focus specifically on recontextualization processes in foregrounding texts (Bray, 2019), that is, the *CB* news article and the tweets mentioning or citing the article. These tweets are written by several authors and institutions, including the main author of the article, other contributing authors, the journal where the article was published, the institutions where the authors work, the author of the *CB* text, and other researchers.

In this case study we analyze recontextualization processes (Van Leeuwen, 2008) and also address visual rhetoric (Tseronis and Forceville, 2017) and aspects of intertextuality and hypertextuality (Bazerman, 2004a; Fairclough, 2003). We draw on Bezemer and Kress (2008), Van Leeuwen (2008), and Luzón (2013) to explore what elements of the source text are moved to a new context, what recontextualization strategies the different rhetors use, and what knowledge transformations take place. Data from frequency lists retrieved with Wordsmith Tools (Scott, 2008) are used to narrow down the discussion of how contents are transformed from one text to another. Intertextuality is analyzed by identifying tangible traces of other texts and looking at what relevant texts/voices are included in the recontextualizing text—e.g. those of the author of the research article, other experts—the form in which they are included—e.g. direct quotation or attribution/mention—and how they are related to each other. We also examine how visuals are recontextualized and contribute to meaning in new multimodal ensembles. Additionally, we comment on features of hypertextuality to determine the function of links in the recontextualizing texts.

Figure 4.5.1 outlines the interrelations between the *NC* research article, the *CB* news article and tweets by various rhetors.

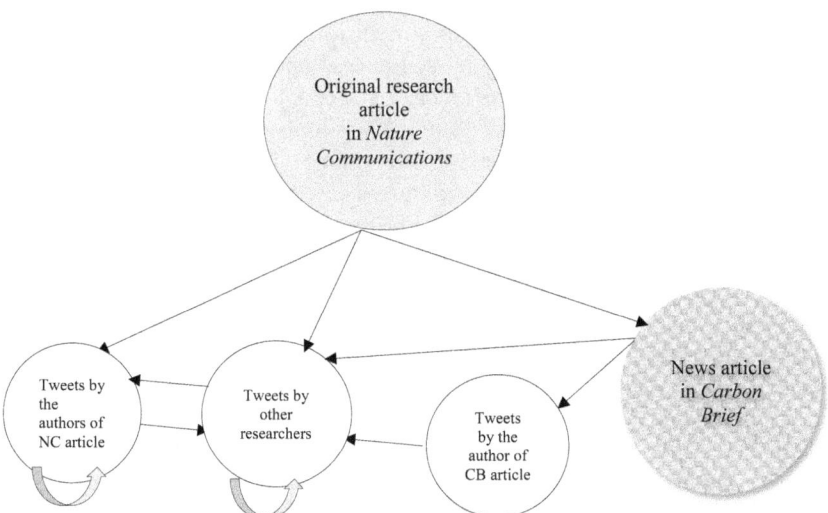

Figure 4.5.1 Interrelations between original article, news article, and tweets.

Recontextualization in the news article

CB defines itself as a "website covering the latest developments in climate science, climate policy and energy policy" which specializes "in clear, data-driven articles and graphics to help improve the understanding of climate change." The CB website features, among other genres, texts that discuss and foreground specific aspects of published research articles, such as the news article written by Tandon (2019). In this news article several strategies are used to recontextualize the NC article to meet the information needs of a wider audience. First, the title of the NC article, which focuses on the phenomenon, is transformed in the title of the CB news article to focus on the consequences, so "Phenological shifts in lake stratification under climate change" becomes "Climate change could cause 'irreversible impacts' to lake ecosystems." Unlike the original NC title, the CB title represents an action—i.e. "cause"—brought about by a non-human agency—i.e. "climate change"—to make clear from the beginning the negative impact of climate change. "Climate change" becomes the subject of a verb ("cause") with negative connotations. The object of the verb "cause," expressed with the collocation "irreversible impacts," occurs three times in the CB article, whereas there is no occurrence in the NC article, which shows the change of focus.

Another recontextualization strategy is the rearrangement of the rhetorical elements of the NC article to meet the purposes of the new text. The rhetorical pattern of the article is adapted to the audience's expectations and needs. The structure of the original article is as follows: Introduction—Results—Discussion (where consequences are discussed)—Methods. The CB text follows a different structure: Main claims—Consequences—Methods—Specific findings. While in the NC article the main claims appear at the end of the introduction and in the conclusions, in the CB text the main claims are foregrounded in the first sentences. The main claim is presented in the CB article in the form of a brief statement ("New research shows that lake 'stratification periods' (...) will last longer in a warmer climate") and then elaborated with more specific claims. The novelty of the main claim is explicitly stated with the evaluative adjective "new." Consequences are foregrounded, first, by placing them immediately after the main claims ("If stratification periods continue to lengthen, we can expect catastrophic changes to some lake ecosystems") and, secondly, by devoting a single section to the consequences, and using the heading "fatal consequences." Neither the word "fatal" nor the word "consequences" occurs in the NC article.

The Methods section is debased and given less value in the CB news article. The NC article presents a very detailed technical description of the research methods, including statistical methods and all the lake datasets used in the study. In contrast, the description of the methods in the CB article is brief, general, and free of technical language, but it provides enough information to help the reader understand the subsequent results and the data presented in the figures and tables. The extract illustrates this point.

In the study, the authors determine historical changes in lake stratification periods using long-term observational data from some of the "best-monitored lakes in the world" and daily simulations from a collection of lake models. They also run simulations of future changes in lake stratification period under three different emission scenarios, to determine how the process could change in the future.

Avoidance of specialized jargon is also observable in the explanation of the tables and figures. Technicisms in the *NC* article ("RCP 2.6, 6.0, and 8.5"—Representative Concentration Pathways) are replaced by more comprehensible terms (e.g. "low emission scenario," "extremely high emissions scenario" *vs* "RCP 2.6," "RPC8.5"), which indicates that the contents of the article are reinterpreted in a more explanatory way.

Processes of recontextualization can also be traced in the selection and transformation of the figures of the original article to meet the purpose of the *CB* text. The *NC* article includes five main figures, which in turn consist of several smaller figures (e.g. "Figure 4a, b, c, d, e, f"), and supplementary information with more figures and tables. To present the results visually, the *CB* text selects four figures from the *NC* article, which focus on specific information (Figures 4e, 5a, b, c). The figures complement the text by extending it. The fragments of text preceding the figures introduce them and describe what data they present (e.g. "The figure below shows the average change in lake stratification days") and the texts following the figures present the conclusion deriving from the results in the figure (e.g. "The plot shows that the average lake stratification period has already lengthened"). Interestingly, the rhetor replaces three figures of the *NC* article (4a, c, e) with a single table, where the complex information in the figures is represented in a way that is easy to understand by the lay public. In addition, the *CB* text also includes an image that does not appear in the original research article published in *NC*. This image is the first element in the document, therefore occupying a very prominent position. Taken from a stock photo collection, the image shows a lake with algal bloom (i.e. accumulation of algae in water, which results in a layer of greenish scum on the surface). Two texts are superimposed on the image: the title of the article ("Climate change could cause 'irreversible impacts' to lake ecosystems") and the image caption ("Algal bloom in Lake Kochelsee in Bavaria"). Thus, the image interacts with the superimposed texts to frame the *CB* article as focusing on the negative consequences of lake stratification. While all the figures in the original research article play a key role in providing evidence for the interpretation (see also Miller, 1998), in the *CB* text this image is added for persuasive purposes. It attracts the reader and complements and illustrates the claim made in the title.

In the *CB* article, strategies such as the explanation of technical terms and concepts tailor information to the assumed knowledge of the general audience. In the following extract, the term "stratification period" is defined, the technical term "phytoplankton" is clarified with a hypernym—"a type of algae"—and the hyperlinked term "trophic mismatch" offers readers further information.

> New research shows that lake "stratification periods"—a seasonal separation of water into layers—will last longer in a warmer climate.
> The study notes that a shift to earlier stratification in spring can also encourage communities of phytoplankton—a type of algae—to grow sooner, and can put them out of sync with the species that rely on them for food. This is called a "trophic mismatch."

Here, plain language is used and recontextualization involves the addition of information that the public needs in order to understand the meaning of technicisms. The term "trophic mismatch" is briefly defined in the *CB* article, but readers are given the choice to access some explanatory knowledge of the concept and its effects through the hyperlink, which leads them to a related article in the journal *Global Change Biology*.

Intertextuality and hypertextuality

The *CB* text incorporates a polyphony of voices that can be traced through features of intertextuality and hypertextuality. To create her text, the author of the *CB* article incorporates information from the *NC* article, but also from interviews with the lead author and other researchers who comment on the research findings, and from other texts through hyperlinking. The frequency lists of the *NC* and *CB* articles reveal a difference in which voices are incorporated in the texts and how. The *NC* research article contains abundant occurrences of "we" (73.06 per 10,000 words), which mainly account for the recurring pattern "we used the" (e.g. "For this comparison, we used climate data from ERA5") and collocations with other action verbs, such as "compared," "selected," "calculated," showing that the researchers make themselves visible when reporting the method. However, in the *CB* news article "we" occurs only once, and it does not refer to the author of the *CB* text, but to the authors of the *NC* article. "We" occurs in a quotation from an interview with the lead author of *NC* ("we can expect catastrophic changes to some lake ecosystems, which may have irreversible impacts on ecological communities the lead author of the study tells in *Carbon Brief*"). This quotation, which includes the evaluative noun phrases "catastrophic changes" and "irreversible impacts," conveys urgency and the need to take action. This contrasts with the objective reporting of the original research article, where the consequences are expressed in a neutral tone, always circumscribed to the research findings:

"We expect that the changes in stratification phenology that we have described here will have far reaching implications." The comparison of the frequency lists of the *NC* and the *CB* articles also reveals that while the former does not contain reporting verbs, the latter contains several reporting verbs ("finds," "tells," "shows," "says," "adds"), three of them occurring with a frequency of at least 5 or more occurrences ($f= \geq 5$): "shows" (9 occurrences), "finds" (8 occurrences), and "tells" (7 occurrences).

Several techniques of intertextual representation can be identified in the *CB* text which show how the rhetors incorporate other texts and voices, and in some cases add information not presented in the *NC* article. Mention of the *NC* article, rather than the authors, is used to report the main knowledge claims with less technical language. Concordance lines of reporting verbs show that noun phrases referring to the original research article (e.g. "new research," "the study") collocate with the verb forms "finds" or "shows" introducing the reporting of findings. The verb forms "tells" (always in the collocation "tells *Carbon Brief*") and "says" are used to introduce direct and indirect quotations of long extracts from interviews with the lead author and with other researchers, rather than from the original article (e.g. "the lead author tells," "O'Reilly says"). Direct interview quotations from the lead author are used to reinforce what is said in the original research article, emphasizing the negative effects of stratification and climate change, and even to add ideas that are not expressed in the original article. The *CB* text also incorporates quotations from other researchers (whom the author refers to as "not involved in the study" to stress their objectivity) to add information supporting or expanding the findings in the *NC* article or to provide other experts' evaluation of the study, enhancing its credibility and relevance.

> O'Reilly tells *Carbon Brief* that "it's clear that these changes will be moving lakes into uncharted territory" and adds that the article "provides a framework for thinking about how much lakes will change under future climate scenarios."

Finally, in the *CB* text the quotation of selected short fragments from both the article and the interview which include evaluative adjectives (underlined for emphasis) is used strategically to foreground and emphasize the negative impact of climate change, as shown in the extracts.

> He tells *Carbon Brief* that the impacts of stratification are "<u>widespread and extensive</u>," and that longer periods of stratification could have "<u>irreversible impacts</u>" on ecosystems
>
> Oxygen depletion can have "<u>fatal</u> consequences for living organisms," according to Dr Bertram Boehrer.

The *CB* text also incorporates other voices/texts through links of different types, which perform different functions. The link to the source article provides credibility to the information in the *CB* text and gives expert readers access to the detailed and technical information in the original text ("The study, published in <u>*Nature Communications*</u>, finds that …"). Links to sites such as Wikipedia where concepts are explained (two links) serve to make these concepts understandable to a wide public and thus make the *CB* text cognitively accessible to this public. There are also links to a variety of specialized and popular sources—articles in journals, other articles in *CB*, news outlets such as *The Guardian*—whose purpose is to extend the readers' knowledge on the negative impacts of climate change. For instance, the hyperlink in "These longer periods of stratification […] can drive toxic <u>algal blooms</u>" leads to a popularizing article in *The Guardian* that provides a definition of algal blooms and explains their harmful effects, illustrating them with pictures. The link in "Fish often migrate to deeper waters (…) for example during a <u>lake heatwave</u>" leads to another article in *CB*, which explains the impact of future lake heatwaves ("Lake heatwaves will be 'hotter and longer' by the end of the century"). Other links establish the credibility of the expert voices incorporated in the text and therefore the reliability of the information. Whenever an expert, including the author, is mentioned or quoted there is a link to his/her web page or Twitter account, where the readers can find more information on their research, and to the institutions where they work (e.g. "<u>Dr Richard Woolway</u> from the <u>European Space Agency</u> is the lead author of the article," "<u>Dr Dominic Vachon</u>—a postdoctoral fellow from the <u>Climate Impacts Research Centre at Umea University</u>").

Recontextualization in Twitter

As shown in Figure 4.5.1, the information in the *NC* article is taken up by various tweeters. Many of their tweets are actually retweets, intended to increase the visibility of the article. The information in the *CB* text is also taken up in tweets, for example by the author of the *CB* text herself and by the National Center for Science Education (NCSE). All these tweets are in turn intertextually—and hypertextually—related to other tweets by making use of several techniques and features: URL hyperlinking, quoting and embedding, retweeting, hashtags, and the @mention feature.

The first author of the *NC* article, Dr Woolway, wrote three tweets to draw attention to the publication. The first one publicizes the article and recontextualizes the main claims (see Figure 4.5.2). The main claims of the article ("we show that climate change during the twenty first century will have a considerable influence on stratification phenology, with lakes across the Northern Hemisphere stratifying sooner and maintaining their stratification for longer. Our results show that the duration of the thermally stratified period will increase within a warming world") are simplified into a single claim where the focus is on the agency, and therefore, the consequences of "warming" ("Warming will cause lakes to stratify sooner and

Figure 4.5.2 Tweet by the lead author of the NC article.

remain stratified longer"). This short message will probably attract expert readers, already familiar with the concept of stratification, and encourage them to access and read the whole article. The tweet also embeds a quote card, a multimodal ensemble composed of short quotations/fragments of the original article, including the beginning of the title, part of a figure, which makes the tweet visually appealing, and a short text which emphasizes the originality and relevance of the article, ("there is little understanding of…"), thus seeking to persuade the readers to access it.

The features of the tweet genre impose some length restrictions. That is why the main claim of the original research article is transformed in the tweet to make it as concise as possible. However, the medium also offers affordances that make it possible to connect the tweet to other texts. For example, the tweet incorporates a link to the article, through which readers can access the full text. The title of the NC article has also been transformed to include three hashtags—#lake, #climate, #change—with words which are among the 20 most frequent keywords in the original research article. The hashtag is a feature of tweets through which the author of the tweets creates intertextual relations with other tweets sharing the same subject matter (Page, 2012). Hashtags contain "searchable language" and thus readers can click on them to find other tweets on the same subject (Zappavigna, 2011). The author of the tweet also incorporates other voices with the @mention feature (e.g. @ISMImpacts, @esaclimate). This feature is used to directly address other users, mention them, or provide access to their accounts. As Zappavigna (2017: 210) puts it, it "is a kind of 'amplified' reference and potential tool for self-promotion."

The other two tweets by the main author are quite similar. They both display an image of a lake, not found in the NC article, showing a valuable natural environment worth protecting (Figure 4.5.3). The image, which has been taken from the database of the European Space Agency, the institution where the lead author works, and also incorporated in the tweet via @

R lestyn Woolway
@riwoolway

#LakeTahoe, seen in the middle of this @esa #Envisat
image, is world famous for its natural beauty. However,
this #lake, and many others, is under threat from
#ClimateChange See our latest research
nature.com/articles/s4146... (image: esa.int)
@esaclimate @sciam #EGU21

Traducir Tweet

Figure 4.5.3 Tweet by the lead author of the *NC* article.

mention, has several functions: it expands knowledge of the information in the text ("Lake Tahoe"), it attracts the audience's attention, and it contributes to engagement and persuasion (see Lazard and Atkinson, 2015). The image reinforces the first idea in the tweet ("world famous for its natural beauty") and represents the current situation, which contrasts with the "predicted future" expressed with language that conveys urgency or risk ("under threat"). Interestingly, although the risk message is prominent in the tweet, the word "threat" only occurs once in the original research article ("[...] as well as provide a serious health threat for cattle, pets, and humans") and there is no explicit reference to the lakes being "under threat," although this threat is implied. The rhetor therefore recontextualizes the implications of the study, by stating them explicitly and concisely, so that it can be easily digested by a diversified audience (one of the mentions is @sciam, the Twitter account of the popular science magazine *Scientific American*).

The Twitter accounts of the journal and of the institutions where the authors work also display tweets that take up the claims of the article. These tweets include mentions of the main author (@riwoolway) that direct readers to his Twitter account, a link to the article, and an embedded picture of the first page of the article, displaying the title, the list of authors and the abstract. Other tweeters—including other contributing authors—refer to the *NC* article by retweeting previous tweets; 96 out of the 106 tweets that mention the article are actually retweets. Tweeters may simply retweet, or they may retweet and comment by embedding a fragment of a previous tweet or commenting on the research, which reveals further processes of transformation and discoursal uptake. In the latter case, the comments always include positive evaluation ("have a look at this great contribution," "what a brilliant work"), which contrasts with the research article, which contains no evaluative positive adjectives. In the tweet below (Figure 4.5.4), for instance, the author retweets one of the lead author's tweets, adds one of the main claims, not included in the embedded tweet ("Longer stratification could have implications for a range of lake processes"), and also her reaction to the research and to the contribution of a group of researchers to which she belongs, namely researchers investigating lake ecosystems at the CEH ("really great to see …"). She uses mentions and hashtags to indicate affiliation and to link to the Twitter of the UKCEH Lake Ecosystems project. The comment explicitly reinforces the impact of global warming with a focus on stratification.

Ellie Mackay @DrEllie_Mackay · 19 abr.   ···
Really great to see @UKCEH_LakeEco #longtermmonitoring sites in the
English #lakedistrict contributing data to this analysis. Longer stratification
could have implications for a range of lake processes.

R Iestyn Woolway @riwoolway · 19 abr.
New paper "Phenological shifts in #lake stratification under #climate
#change Warming will cause lakes to stratify sooner and remain
stratified for longer nature.com/articles/s4146... @ISIMIPImpacts
@esaclimate @cfes_dkit @mariescurie_ire #EGU21

Figure 4.5.4 Tweet with comment.

Some tweets about the original research article did not take up information directly from the article, but from the *CB* news item, which functions as an intermediary genre. An example is a thread of four tweets written by the author of the *CB* text, which summarizes and foregrounds the most important information in the *CB* article. Manifest intertextuality can be traced in the first tweet (Figure 4.5.5), which quotes the definition of lake stratification period, the fragment of the *CB* article which presents the main finding ("Lake stratification periods will last longer in a warmer climate"), and the main consequence/implication. It also embeds the title of the *CB* article, mentions (@AyesaTandon, @riwoolway), and the first picture appearing in the *CB* article, which, as pointed out above, reinforces the negative consequences. The second tweet (Figure 4.5.6) gives a brief summary of the

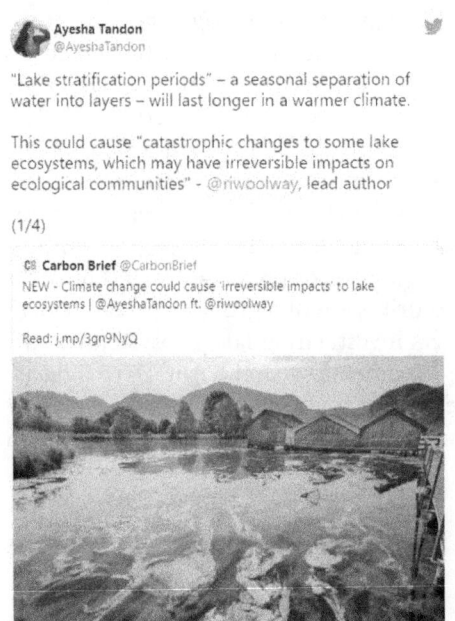

Figure 4.5.5 First tweet by the author of the *CB* text.

results, including Figure 4a from the original article, which also appears in the *CB* text, to reinforce what is presented in the text. The third tweet (Figure 4.5.6) lists the specific consequences of the stratification period, and the last tweet presents the drives of longer lake stratification periods (rising air temperatures, and slower surface wind speeds). Therefore, in four tweets, the thread presents a brief concise summary of the research, easily understood by a wide public. Listing and emojis (Figure 4.5.7) help to organize this summary clearly, with emojis thus functioning as visual metadiscourse (see De Groot et al., 2015). Importantly, tweets 2–4 cannot be understood in isolation, but only as part of the summary thread, revealing a type of connection between texts that makes it possible to overcome the length constraints of the tweet genre.

Interpretation

This case study has shown processes of discoursal uptake and recontextualization across genres and explored how both text and visuals travel from a research article to other genres where they are recontextualized and repurposed. As commented earlier, fragments of texts and visuals are extracted from the original situational context—an online research article—and reframed in new contexts, namely online science news and Twitter.

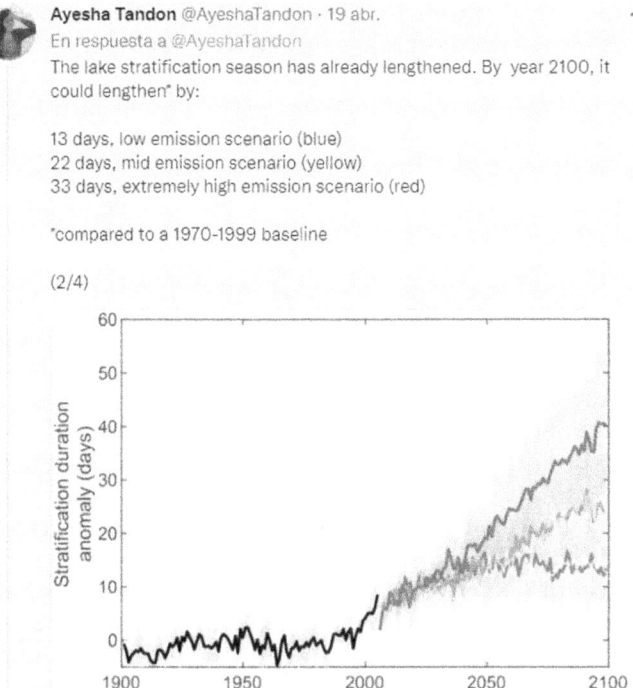

Ayesha Tandon @AyeshaTandon · 19 abr.
En respuesta a @AyeshaTandon
The lake stratification season has already lengthened. By year 2100, it could lengthen* by:

13 days, low emission scenario (blue)
22 days, mid emission scenario (yellow)
33 days, extremely high emission scenario (red)

*compared to a 1970-1999 baseline

(2/4)

Figure 4.5.6 Second tweet by the author of the *CB* text.

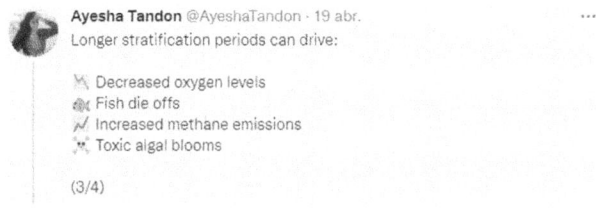

Ayesha Tandon @AyeshaTandon · 19 abr.
Longer stratification periods can drive:

Decreased oxygen levels
Fish die offs
Increased methane emissions
Toxic algal blooms

(3/4)

Figure 4.5.7 Third tweet by the author of the *CB* text.

Through processes of discoursal uptake, the contents of the original text are reframed to make it relevant for specific publics or to achieve other communicative purposes. When the contents of an article are re-entextualized in Twitter, they are reframed in several ways depending on the rhetors' intentions, which may range from promoting, spreading, and giving visibility to the contents of the original text, to showing support to the findings and affiliation with a research community or disseminating science and improving the public's scientific literacy.

As seen in this case study, the reframing and reinterpretation of the information from the original research article in the *CB* article and in the tweets involves processes of selection, rearrangement, transformation, and addition. Specific information and visuals are selected to be moved to the new context, taking into account the rhetor's intentions, the intended audience, and the affordances and constraints of the new genre. This process of selection is of particular relevance in highly space-constrained genres, such as Twitter, where the reduced number of characters makes it necessary to select and combine semiotic resources skillfully to achieve the intended communicative goal. In the *CB* text, specific information is rearranged by placing the main claim of the *NC* article and the consequences at the beginning of the *CB* article, thus foregrounding them. Transformation of meaning-making elements also serves to meet the interests and communicative needs of a different "epistemic community" (Van Dijk, 2014: 4): titles are transformed to emphasize the agency of "climate change," and the figures in the original article are transformed into tables, to make them cognitively accessible to a wider audience. Addition strategies also play an important role in the recontextualization of the *NC* article in the other texts. They involve addition of content and visuals from other sources, elaboration of content through hyperlinks, and addition of evaluation or reaction, particularly in tweets. The *CB* news article has been constructed by uptaking contents and visuals from the *NC* article, and adding other texts, voices, and visuals. The added information—i.e. from interviews with the lead author and researchers—is used to legitimize some aspects reported in the text—i.e. the negative effects of lake stratification—or to provide more detail. Content is elaborated through hyperlinks that provide extra information that might help the readers to understand or that help the rhetor to "manipulate" the readers and guide them to a specific stance. Therefore, recontextualization involves the orchestration of various semiotic resources—taken both from the original article and from other texts—to achieve the purposes of the new genres.

This case study has also shown the important role visuals play in the recontextualization of the original article in the news article and in the tweets. In addition to selected visuals from the *NC* article, the rhetors also added other visuals that helped them achieve particular rhetorical purposes (see Tseronis and Forceville, 2017). The two figures transported from the *NC* article as well as the table based on its figures provide visual support for the claims in the text, thus complementing it, and increase the credibility of the rhetor, since they are credited to the author of the original article. The "added" pictures—e.g. the picture of the effects of algal bloom on lakes in the online news article or pictures of the lakes in the tweets, taken from sources other than the original article—have a persuasive effect. They are used to engage the viewers affectively and prompt a feeling of concern about the environment. They occupy a prominent position in the article and in the tweets, and frame the text, guiding the readers to the rhetor's perspective.

In this case study we have also shown how all the online genre texts forming part of a genre network online are connected through various types of intertextual and hypertextual links. By tracing intertextuality, we have identified how the texts incorporate extracts of other texts in several ways— by responding to previous texts, mentioning them, quoting extracts directly or indirectly, or embedding fragments from other texts. The hypertextual affordances of the medium expand the possibilities for intertextuality and blur the boundaries between genres connected online. The connection to other texts through hyperlinks serves several purposes in the dissemination of research online, namely, to provide credibility to the information, give expert readers access to technical information, make a text more cognitively accessible for the lay public, or enhance the audience's knowledge about an issue and persuade them of the author's stance, as seen with the links to texts that focus on the negative effects of climate change.

The study has further shown that the technological affordances and features of Twitter —i.e. @mention, hashtags, retweeting, threads—make it possible to enhance connectivity and create links with a large number of texts (for further discussion see Page, 2012; Zappavigna, 2011, 2017). All these connected texts therefore offer various paths to find the information appropriate to readers' communicative needs and interests. The space-constrained nature of tweets and their promotional and attention-getting nature may explain the importance of connectivity in this genre. We have seen that the features of Twitter enable new forms of intertextuality not possible among pre-digital genres. Hashtags serve as access points for finding related tweets on a single topic, which may in turn provide access to other texts. The possibility of replying and mentioning facilitates interaction and creates new relations between texts. Retweeting "quotes" a tweet (or part of it) to endorse it, or comment on it, and make it widely visible. This case study has suggested that the affordance of retweeting for "message rebroadcasting" has an important role in the dissemination of science online. By transforming the article's claims to be presented in a tweet, the author can broadcast them not only to the disciplinary audience, but also to other audiences, which include experts in other disciplines and the interested public. The other authors of the article (and other stakeholders) uptake tweets reporting on the research and help to broadcast this research by commenting or retweeting. Unlike the relations between a research article and other articles (or genres) intended for the disciplinary community, the relations between an article and the social media—tweets—that uptake it bridge the boundaries between the disciplinary community and other audiences. Tweets and retweets of articles can thus help to improve awareness of and access to research among the interested public, which is of high importance in the case of research related to climate change.

To conclude, this case unveils complex and dynamic aspects of discourse construction and interpretation in online connected genres. The polyphony of voices that results from the uptake and recontextualization of the research article in subsequent genres, and the orchestration of various semiotic resources—text, visuals, hyperlinks—creates multiple meanings, which are of relevance to various audiences. Readers can move from the factual report of the phenomenon in the research article to the description of the effects of climate change, emphasized by the voice of experts in the news article, or they can move from the concise presentation of the main results in the tweets to the detailed description in the research article or the news article.

Notes

1 UNFCCC. The Paris Agreement. United Nations Framework Convention on Climate Change (2015). Available at http://unfccc.int/paris_agreement/items /9485.php. Accessed 16 Oct 2021. Video also available on YouTube at https://www.youtube.com/watch?v=5THr3bFj8Z4&t=80s
2 Mengel, M., Nauels, A., Rogelj, J. et al. Committed sea-level rise under the Paris Agreement and the legacy of delayed mitigation action. *Nat Commun* 9, 601 (2018). https://doi.org/10.1038/s41467-018-02985-8
3 Rising to the challenge of surging seas. *Nat Commun* 8, 16127 (2017). https://doi.org/10.1038/ncomms16127
4 "What happened last time it was as warm as it's going to get later this century" by Howard Lee https://skepticalscience.com/news.php?n=4176 a repost from https://arstechnica.com/science/2018/06/are-past-climates-telling-us-were-missing-something/
5 http://ucrel.lancs.ac.uk/llwizard.html
6 Fletcher, W. H., M. Jalo, and Laine, A.-L. (2021). The effect of host community functional traits on plant disease risk varies along an elevational gradient. 10:e67340 DOI:10.7554/eLife.67340 Licensed under a Creative Commons Attribution 4.0 International License.
7 Elsevier defines it as a statement that "should be written to a broad audience, at an undergraduate level, and limited to 120 words" (https://www.journals.elsevier.com/acta-biomaterialia/news/the-statement-of-significance) [Last accessed 3 October 2021].
8 Millet et al. 2019.
9 http:// www.jove.com
10 https://www.zooniverse.org/projects/tokehoye/pollinatorwatch. We wish to thank the PollinatorWatch research team for granting us permission to reproduce text and images from their project homepage.
11 Woolway, et al. 2021.
12 Tandon 2019.

5 Towards complex digital communication models

In this book we have explored networks of genres in digital science communication, focusing on processes of scientific knowledge recontextualization and intersemiotic relations across and between genres and modes. It was our aim to demonstrate how genre analysis is a key theoretical and interpretative framework for analyzing the interrelations between connected genres in online environments and for gaining a better understanding of today's complex models of digital science communication. With the case study research, we have sought to cover genres that are representative of the professional sphere of (scientific) communication and para-scientific genres belonging to the sphere of public communication of science (Kelly, 2014; Kelly and Miller, 2016). Moreover, the selection of genre exemplars has also addressed diversity in terms of hypermodality affordances. We would like to stress once again that in each case study the very nature of each genre network, and the semiotic modes involved, deemed it appropriate to analyze in detail one particular layer of textual analysis over the others, or to focus on one semiotic mode over the other modes. By this means we have also aimed to illustrate a range of processes of knowledge transformation, refocusing, reinterpretation, and resemiotization.

With the development of Web 2.0 there has been an unprecedented proliferation—and speciation—of genres written, spoken, and both written and spoken, as well as new modes and media for communicating science online. It would have been overambitious to include such generic diversity in a single book. Notwithstanding this limitation regarding generic representativeness, our main goal was to move forward with the claim that in order to understand science communication on the web, textual analysis needs to include *dynamic aspects of discourse construction* (our emphasis). We have taken two new directions in this respect. The first is the exploration of dynamic aspects of discourse construction, not in print genres, but in web-based connected genres. Here, we have applied Bhatia's critical genre analysis framework to describe intersemiotic relations among and between different genres and modes. The second direction is methodological. Bhatia's framework was not fully clarifying or, to be more precise, not sufficiently methodologically

DOI: 10.4324/9781003147732-5

robust to empirically analyze genres in digital media, for example, regarding the procedures or steps to be followed for tracing genre connections and multisemiotic meaning making, as both are inherent features of genre networks in such environments. With the aim of clarifying the dynamic aspects of discourse construction, we have shown that when analyzing the complexity of web-based forms of genre collectivity, we need to pan and zoom simultaneously. Panning allows us to capture the whole picture of the processes of sequential uptake and recontextualization among and between genres, and even within genres, in the case of macro genres in web portals. At the same time, zooming allows us to magnify the salient features of the different textual layers of the genres of each network, bringing into close-up the hybridity, interdiscursivity, embedding, generic colonization, and evolutionary development of genres, among other generic phenomena. Here, it should be acknowledged that the analysis of the small-scale corpora used for each case study might be regarded as somewhat limited because it only portrays relatively small networks of genres. Indeed, more complex networks, i.e. networks involving a higher number of genres, could have been included. The analytical decisions taken to explore different textual layers of the genres forming the network and aspects of multimodality encouraged us to select networks of a manageable size.

Pedagogical applications

There are a number of interesting and timely applications of the case study research. The networks of genres analyzed, along with other genre networks that the teacher of Languages for Academic Purposes and English for Academic Purposes and (LAP/ESP) may find it appropriate to work with in the classroom to better address students' language and communication learning needs, can support academic literacy learning processes, above all at graduate and PhD levels. Together with these early career researchers, experienced researchers interested in enhancing their academic and digital skills in professional and public communication of science may also find it useful to get to know the complex dynamics of digital science communication.

Taken as models of science communication, networks of genres online can be used in formal instruction to help students learn how to express their worldviews and perspectives taking into account that the texts they produce may involve multisemiotic meaning making and may be connected (hyperlinked) to one or more texts for meaning enhancement. Exemplars of genre networks can familiarize students with the expanding genre diversity in today's digital era and reveal the variety of genre connections and intersemiotic relations—e.g. intertextual, multimodal, and hypertextual—in the online environment. Furthermore, exploring genre networks in formal learning can also pave the way to becoming acquainted with issues related to the context of academic knowledge production, co-construction, and negotiation among peer scientists (as shown in case studies 2 and 3) and between scientists and other science stakeholders, including the general public (as illustrated in the remaining case studies).

Exemplars of genre networks in digital environments can be a source of rich linguistic input to trigger language acquisition and learning processes in EAP/LAP contexts. All the networks selected for the case study research contain authentic texts of science communication online. Pedagogical activities drawing upon these real texts can provide analytical skills practice, specifically helping students to recognize or notice genre interrelations and processes of knowledge recontextualization and discoursal uptake. Activities involving the analysis of genres working in conjunction with other genres can also elicit critical discussion of the situational and sociocultural contexts that shape the form and substance of genres, as well as their dynamism and openness. Bawarshi and Reiff (2010: 50) explicitly underline the pedagogical value of attending to genre networks to understand not only single genres but also "how genres interact with one another in complex ways to achieve dynamic purposes." Developing critical reading skills can also be supported through reading activities and analytical tasks that involve the identification of "rhetorical agency" (Artemeva, 2004: 4), and the range of perspectives or polyphony of voices that a given genre network may embrace.

Examples of genre networks can also be integrated in Data-Driven-Learning (DDL) activities (Charles, 2015; Johns, 1991; Tribble, 2001). Formal instruction can engage students in the use of free software and corpus tools, for example to find out the reasons why some genres exhibit a higher degree of conceptual depth and syntactic complexity than the remaining genres forming the network. Tasks based on data from genre networks can also illustrate the formation of colonies of genres and their shared communicative purposes. Analysis of genre networks can also raise students' awareness of the lexical profile of the genres forming the network and of the network as a whole. Activities can involve learning keywords, as identified in frequency words or lexical profiles, and analyzing how the key content words are used in context in the different texts and how meanings are conveyed in different ways and through different modes. By way of illustration, students can compare the word clouds generated with Atlas .ti or #LancsBox, as shown in Figures 5.1 and 5.2. Comparing these clouds is a task that does not require much cognitive effort and has the advantage of making students aware of how global threats such as climate change are addressed in the different genres of the network (Figure 5.1). Comparing the two figures can also be helpful to understand the outcome of refocusing and repurposing processes in text trajectories. Compared to the previous figure, Figure 5.2 shows how in the original research article of this case study, the natural phenomenon ("sea level") is the most salient semantic meaning, followed by its subsequent effect, "rise," and consequence, "net-zero emission scenarios." Climate change is not the prominent issue under discussion. In the original research article, the rhetors have refocused the topic to give

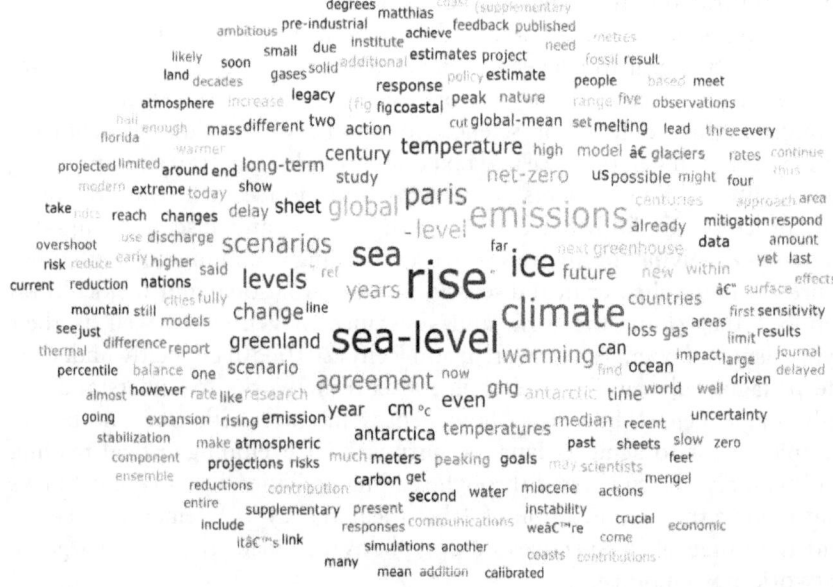

Figure 5.1 Semantic salience in the entire ecology in Case study 1.

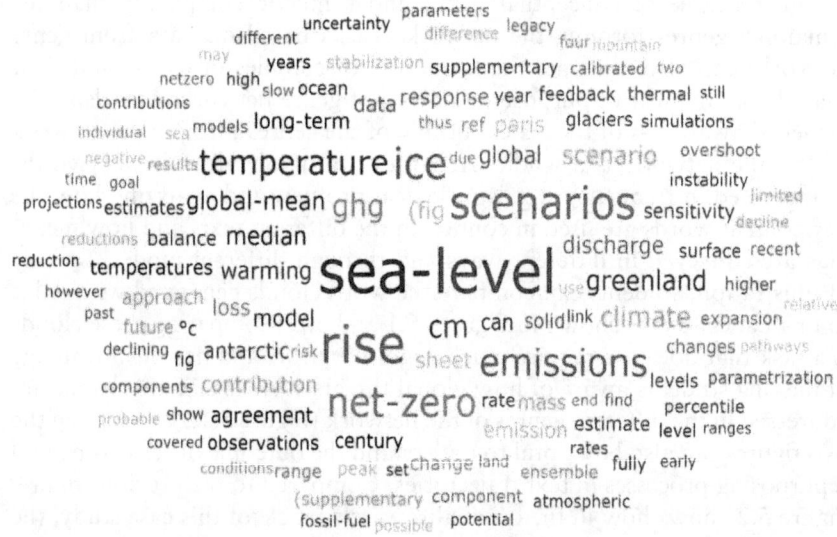

Figure 5.2 Semantic salience of the research article in Case study 1.

greater relevance to the possible future outcomes and evaluate them from a scientific point of view.

Tasks using examples of genre networks online can also engage students in the analysis of the recontextualization processes traced in different textual layers: semantics, grammar and syntax, pragmatics, intertextuality, and other discourse features. For graduate students and early career researchers, it may be useful to pay close attention to discoursal uptake and co-construction of meanings in the sequence of genres illustrated in Case study 2, or to the transformation of scientific contents textualized in expert genres into accessible contents textualized in para-scientific genres such as those discussed in case studies 1 and 5—i.e. research articles repurposed in a lay summary and in tweets and retweets used by researchers to disseminate their article in social media networks. For example, a guided exercise involving cross-genre comparison of the lexical profiles and language patterns—e.g. collocations, words in context, grammar structures, and so on—can prompt an exploration of the interrelations among connected genres at different textual levels. This exercise could also pave the way to raising students' awareness of functional language variation—i.e. register— and the way texts are manipulated by different rhetors to fulfill specific rhetorical goals, for instance, by reformulating information to make it appealing to audiences other than the audience of the recontextualized text.

A comparative analysis of the genres forming a given network can also boost students' understanding of the pragmatics of the texts in relation to their situational context and their socio-historical context. For example, case studies 1 and 5 can serve to compare the perspective of a journal article written by scientists with the perspective of other science stakeholders (institutions, journal editors) and science journalists. Both cases can also illustrate different clines of authorial positioning, reflecting different viewpoints and ideologies. The genre network analyzed in Case study 1 showcases how and why evaluation is conveyed in the texts through epistemic and deontic modality markers such as "can" and "may." This network is also a good example to encourage students to think critically about textual silences, for instance the absence of the modal verb "might," which did not occur in any of the genres of this network, and which explains the rhetors' assertiveness. Tracking the presence and absence of linguistic elements is important for understanding the rhetors' strategic use of language and different argument construction across the texts to respond to different communicative goals.

The previous analytical tasks should necessarily be followed by production tasks that offer the students opportunities to apply the knowledge they have acquired (Swales and Feak, 2009). The pedagogical goal of these language production activities should be to help early career researchers and scientists become more effective communicators of science in and for society. From the explorations of the case studies, a number of learning outcomes can be proposed. Writing tasks could be useful for the students to practice processes of discoursal uptake and recontextualization, for instance, transforming traditional genres such as abstracts and articles into new digital genres such as impact statements, lay summaries, tweets, or citizen science projects, to name a few. Taking case studies 2 and 4 as models for language production, tasks can also prompt practice in writing for multiple audiences and writing for different communicative goals. This practice should involve transforming contents, refocusing and repurposing them to reach different audiences. Specifically, it can engage students in reflective writing, for example, in taking decisions as to what strategies (e.g. linking contents through intertextual references, restructuring arguments, clarifying specialized contents, expanding contents, or simplifying them) are appropriate when transforming one genre into other genres and why. By way of illustration, writing an impact statement and a digest using a journal article as the source text or, as another example, writing a tweet to disseminate a journal article, also paying attention to aspects such as the fixity of the interface design, text modularity, or character-length constraints, among other aspects, could support a reflective approach to writing. Other suggestions provided in the current literature (Pérez-Llantada, 2021b; Gimenez and Roque Gutierrez, 2016) are also worth considering.

Suggestions for digital composing practice such as those sketched out above should necessarily involve practice of knowledge transformation and discoursal uptake across modes and media. Here, once again, reflective writing tasks can be developed to elicit reflection on ways of resemiotizing the information from one mode to another mode, focusing on intersemiotic relations across and between the visual and verbal modes. An example of a side-by-side view of an article and its related visuals available on some journal platforms can be a good start for practicing multimodal composing. By way of illustration, Figure 5.3 shows a side-by-side view of the online

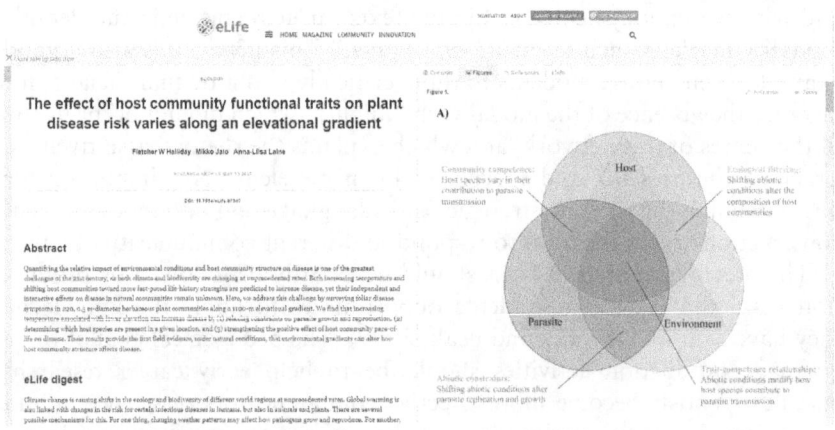

Figure 5.3 Model of a text-visual side-by-side view for prompting production practice.

article analyzed in Case study 1, which can be used as a model for students to create one or several visual elements and gain skills in making both modes either complementary, concurrent, or divergent, following Jones and Hafner's (2012) general taxonomy. This writing practice could also reinforce students' awareness of the preferred navigation options (namely, whether accessing the text first or the visuals first) that their intended readership may follow. More importantly, it can raise awareness of how the intersemiotic relations established influence the way information is provided in the two modes so that contents are fully understood.

More cognitively demanding tasks could involve practice in composing video methods articles associated with research articles. The exemplars explored in Case study 3 could be useful models for students to explore and understand the intersemiosis of coexisting modes of meaning making to create, say, a short video clip showing the research methods that they use and to report such methods in an associated journal article. Other multimodal composing tasks can initiate the students in communication for public engagement in science. Case study 4 can be taken as a model of internally hyperlinked genre embedding and as an example of colonies of texts that rely heavily on visual elements (images above all) for the purposes of informing, instructing, and guiding citizen scientists in the data classification process. This writing activity could further elicit students' sensitivity towards the use of digital genres that support and promote the societal impact of scientific research today.

In light of the explorations of the case study research, the creation of multimodal genres should offer practice in ways of integrating different types of audio and visual modes—ranging from still images, moving images, gestures, spoken language, music, and non-verbal sound (see Tseronis and Forceville, 2017)—with written language, by this means forming complex multimodal ensembles. YouTube video clips, podcasts, interviews, audio slide presentations, three-minute thesis presentations, or vlogs, all of them genres that graduate and PhD students in the sciences field might find it useful to compose to give visibility to their research work, could also be easily integrated in these digital multimodal composing tasks.

Implications and future directions

Throughout this book we have highlighted the importance of examining complex types of genre collectivity to obtain a more in-depth view of interactions across genres and modes in digital environments. The bottom-up, text-first approach to genres that we have used, reinforced by other analytical procedures, has uncovered text-internal features of language and convergence (or divergence) of language patterns across the genres, by this means triangulating generic action and intersemiotic relations. In addition, this approach has allowed us to understand multimodal recontextualization practices in relation to contextual aspects, above all those related to the participant relationships and the ideologies underlying discourse construction. On the basis of the evidence found in the case study research, we would argue that the cases mainly contribute to supporting the claim that genres should be viewed in conjunction with other genres to better understand the multilayered connections that can be established between and among them in digital environments. The case study research also lends credence to the importance of examining multisemiotic meaning-making and recontextualization processes "to place individual genres within a heuristically valuable wider context" (Swales, 2004: 23). The intertextual, interdiscursive, and hypertextual dimensions traced within the genre networks that we have examined support Briggs and Bauman's (1992: 132) claim that intertextuality is a prototypical feature of genres that determines their "textual open-endedness." Here, an ecological view of genres in digital environments is relevant and timely too in the digital era (for further discussion see Casper, 2016; Harmon, 2019; Pérez-Llantada, 2019). Such a view foregrounds the generic openness and the dynamic relations between materially linked genres in digital environments that we postulate.

One important advantage of setting up a comparative analysis—i.e. comparing genres to one another by identifying their similarities and differences across textual layers—is that it allows a better understanding of the processes of knowledge (re-)entextualization in text trajectories and of the way the same rhetors—or, in certain networks, different rhetors—construct meanings to accomplish social exigences and "private intentions" (Miller, 1984: 164). A comparative approach further supports the ecological view of genres as open constructs that can be either internally or externally hyperlinked genres. Looking at the wider context of genres and tracing processes of knowledge transformation and recontextualization has also proved useful to uncover how rhetors exploit or manipulate generic resources in various ways creating, as Bhatia (2004: 156) notes, "hybrid genres, which may include mixing, embedding, bending and appropriating of generic resources." In the case study research, we have found textual evidence, at least *prima facie*, to be able to claim that genres online are hybrid. As illustrated in Case study 2, the distinct features of the impact statement compared to those of its connected genres suggest that this genre has borrowed features of other emerging digital genres in online journal publications, such as the five-bullet point research highlights appearing in some journals, that "capture the results of the research as well as new methods that were used during the study (if any)."[1] As shown in this case study, the impact statement synthesizes the essence of the article, though instead of referring to methods, it underlines results and implications. It also shares the same communicative purpose as that of the editors' short text accompanying the *Nature Communications* article. These comparative views also make it possible to affirm that new media environments invite the creation of hybrid genres with formal and generic similarities with their associated genres, but at the same time with distinct communicative goals and subject to different rhetorical exigences.

Methodologically, the incorporation of text-external factors in the analysis of text-internal features allows a contextualized treatment of issues of social concern and a complex, web-mediated polyphonic "conversation" on such issues. This conversation is, as shown in the case study research, socio-historically situated. In the cases analyzed, global concerns such as climate change or health issues are approached from multiple perspectives and target multiple intended audiences. The inherent complexity of genre networks thus requires the implementation of strong analytical approaches to examine knowledge transformation, recontextualization, and resemiotization. On the evidence found in the case studies, we reaffirm the importance of examining genres in conjunction with other genres to further our understanding of the multifaceted nature of genre networks. This analytical endeavor is certainly challenging. We have seen, for instance, that across the case studies analyzed it may not always be possible to systematically apply similar cut-off points or frequency ranges, as genre networks are extremely diverse. We therefore believe that in future research the exploratory analytical approach that we have illustrated will benefit from methodological refinements for the practical analysis of genre networks and more complex forms of genre collectivity that emerge in online environments. For instance, the inclusion of a wider variety of genres, for example spoken genres and audiovisual narratives such as those analyzed in case studies 1, 3, and 4, and other emerging genres such as podcasts, audio slides, or vlogs would add further complexity.

Achieving methodological robustness is even more challenging when we engage in the comparative analysis of mini genres—as those of the macro genre explored in Case study 4—or in the analysis of the re-entextualization of scientific texts into very short mini genres such as impact statements or tweets. As also noted in the case study research, descriptive statistical measures such as frequencies or collocational analysis may not always be a suitable starting point to trace intertextual links and features of interdiscursivity. We therefore need to explore further other layers of the text in order to empirically address language variation. Given these considerations, future research should further enquire into this relatively unchartered territory to offer feasible ways of providing thick descriptions of language and semiotic resources, both in small-scale and large-scale genre networks. Also, because in the future Internet technologies will keep on developing further affordances for its users, it is likely that we will need to rely on more wide-ranging multidisciplinary research to engage in this investigative endeavor. On the web, the textual boundaries shaping the generic integrity of each genre of the network are porous and permeable, hence the importance of advancing theories and methods to understand how multimodal resources such as language, visuals, and aural narratives, *inter alia*, are orchestrated in networks of connected genres.

Pedagogically, we think that it is very important to implement creative ways of teaching and learning about connected genres in order to help students learn key aspects of language, register, and mode variation in (re-) entextualization processes while developing their critical thinking towards the broader contextual aspects of science communication in relation to the Open Science agenda. To date, we do not have any empirical evidence proving the value of the pedagogical applications proposed earlier in this chapter. However, observations from our teaching practice strongly suggest that introducing genre networks has a number of benefits, namely rich linguistic input, exposure to real texts for the (critical) analysis of semiotic modes, and subsequent practice in the recontextualization of scientific knowledge drawing on language and audiovisual elements. We therefore invite the development of empirical studies and reflective practice on teaching/learning approaches along these lines.

Note

1 Source: Elsevier https://www.elsevier.com/journals/journal-of-english-for-academic-purposes/1475-1585/guide-for-authors

References

Aalbersberg, I. J., Heeman, F., Koers, H., and Zudilova-Seinstra, E. (2012). Elsevier's article of the future. Enhancing the user experience and integrating data through applications. *Insights: The UKSG Journal*, 25(1), 33–43.

Adam, C., and Artemeva, N. (2002). Writing instruction in English for Academic Purposes (EAP) classes: Introducing second language learners to the academic community. In A. Johns, ed., *Genre in the Classroom: Multiple Perspectives* (pp. 179–196). Mahwah: Erlbaum.

Adami, E. (2014). Retweeting, reposting, repining, reshaping identities online: Towards a social semiotic multimodal analysis of digital remediation. *Lingue e letterature d'Oriente e d'Occidente*, 3, 223–243.

Andersen, T. H., and Van Leeuwen, T. J. (2017). Genre crash: The case of online shopping. *Discourse Context & Media*, 20, 191–203.

Anderson, P. (2007). What is web 2.0? Ideas, technologies and implications for education. *JISC Technology and Standards Watch, Feb. 2007*. Bristol: JISC. Retrieved from http://www.jisc.ac.uk/media/documents/techwatch/tsw0701b.pdf [Last accessed on 15 July 2022]

Androutsopoulos, J. (2011). From variation to heteroglossia in the study of computer-mediated discourse. In C. Thurlow and K. Mroczek, eds., *Digital Discourse: Language and the New Media*. New York: Oxford University Press, pp. 277–98.

Androutsopoulos, J., and Tereick, J. (2016). Youtube. Language and discourse practices in participatory culture. In A. Georgakopoulou and T. Spilioti, eds., *The Routledge Handbook of Language and Digital Communication*. London: Routledge, pp. 354–370.

Anthony, L. (2019). *AntConc (Version 3.5.8) [Computer Software]*. Tokyo, Japan: Waseda University. Available from: http://www.laurenceanthony.net/software/antconc/ [Last accessed on 15 November 2021]

Anthony, L., and Hardaker, C. (2017). *FireAnt (Version 1.1.4) [Computer Software]*. Tokyo: Waseda University. Available from: http://www.laurenceanthony.net/software [Last accessed on 15 November 2021]

Artemeva, N. (2004). Key concepts in rhetorical genre studies: An overview. *Technostyle*, 20(1), 3–38.

Askehave, I., and Swales, J. M. (2001). Genre identification and communicative purpose: A problem and a possible solution. *Applied Linguistics*, 22(2), 195–212.

Baker, M. (2016). 1,500 scientists lift the lid on reproducibility. *Nature*, 533(7604), 452–454.

Bakhtin, M. M. (1981). *The Dialogic Imagination. Four Essays.* C. Emerson and M. Holquist (trans.), M. Holquist, ed. Austin: University of Texas Press.

Bakhtin, M. M. (1986). *Speech Genres and Other Late Essays.* V. W. McGee (trans.), C. Emerson and M. Holquist, eds. Austin: University of Texas Press.

Baldry, A., and Thibault, P. J. (2006). *Multimodal Transcription and Text Analysis: A. Multimedia Toolkit and Coursebook.* London/Oakville, Equinox.

Ball, Ch. E. (2016). The shifting genres of scholarly multimedia: Webtexts as innovation. *The Journal of Media Innovations*, 3(2), 52–71.

Ball, D. (2016). *The Impact of Open Science.* Retrieved from https://www.fosteropenscience.eu/content/impact-open-science [Last accessed on 1 July 2020].

Bar-Ilan, J. (2005). What do we know about links and linking? A framework for studying links in academic environments. *Information Processing & Management*, 42, 973–986.

Bartling, S., and Friesike, S. (2014). Towards another scientific revolution. In S. Bartling and S. Friesike, eds., *Opening Science.* Cham: Springer, pp. 3–15.

Bateman, J. A. (2008). *Multimodality and Genre.* Basingstoke: Palgrave Macmillan.

Bateman, J. A. (2014). Genre in the age of multimodality: Some conceptual refinements for practical analysis. In P. Evangelisti-Allori, V. K. Bhatia, and J. A. Bateman, eds., *Evolution in Genre: Emergence, Variation, Multimodality.* Frankfurt am Mainz: Peter Lang, pp. 237–269.

Bawarshi, A. S. (2010). Taking up multiple discursive resources in U.S. college composition. In B. Horner, ed., *Cross-language Relations in Composition.* Carbondale: Southern Illinois University Press, pp. 196–203.

Bawarshi, A. S., and Reiff, M. J. (2010). *Genre. An Introduction to History, Theory, Research and Pedagogy.* 1st edition. West Lafayette: Parlor Press.

Bazerman, C. (1994). Systems of genres and the enactment of social intentions. In A. Freedman and P. Medway, eds., *Genre and the New Rhetoric.* London: Taylor and Francis, pp. 79–101.

Bazerman, C. (1997). The life of genre, the life in the classroom. In W. Bishop and H. Ostrom, eds., *Genre and Writing: Issues, Arguments, Alternatives.* Portsmouth: Boynton/Cook, pp. 19–26.

Bazerman, C. (2004a). Intertextuality: How texts rely on other texts. In C. Bazerman and P. Prior, eds., *What Writing Does and How It Does It: An Introduction to Analysing Texts and Textual Practices.* New York: Taylor and Francis, pp. 83–96.

Bazerman, C. (2004b). Speech acts, genres, and activity systems: How texts organize activity and people. In C. Bazerman and P. Prior, eds., *What Writing Does and How It Does It: An Introduction to Analysing Texts and Textual Practices.* New York: Taylor and Francis, pp. 309–339.

Bernstein, B. (1996). *Pedagogy, Symbolic Control and Identity. Theory, Research, Critique.* London: Taylor and Francis.

Bezemer, J., and Kress, G. (2008). Writing in multimodal texts: A social semiotic account of designs for learning. *Written Communication*, 25(2), 166–195.

Bezemer, J., and Kress, G. (2017). Continuity and change: Semiotic relations across multimodal texts in surgical education. *Text & Talk*, 37(4), 509–530.

Bezemer, J., and Mavers, D. (2011). Multimodal transcription as academic practice: A social semiotic perspective. *International Journal of Social Research Methodology*, 14(3), 191–207.

Bhatia, V. K. (1993). *Analysing Genre Language Use in Professional Settings*. London: Longman.

Bhatia, V. K. (1996). Methodological issues in genre analysis. *Hermes: Journal of Linguistics*, 16, 39–59.

Bhatia, V. K. (2004). *Worlds of Written Discourse A Genre-Based View*. London: Continuum International.

Bhatia, V. K. (2008). Towards critical discourse analysis. In V. K. Bhatia, J. Flowerdew, and R. H. Jones, eds., *Advances in Discourse Studies*. London/New York: Routledge, pp. 166–177.

Bhatia, V. K. (2017a). *Critical Genre Analysis: Investigating Interdiscursive Performance in Professional Practice*. London/New York: Routledge.

Bhatia, V. K. (2017b). Methodological issues in genre analysis. *HERMES: Journal of Language and Communication in Business*, 9(16), 39–59. https://doi.org/10.7146/hjlcb.v9i16.25383

Biber, D., and Conrad, S. (2020). *Register, Genre and Style*. 2nd edition. Cambridge: Cambridge University Press.

Biber, D., and Gray, B. (2010). Challenging stereotypes about academic writing: Complexity, elaboration, explicitness. *Journal of English for Academic Purposes*, 9, 2–20.

Biber, D., and Gray, B. (2016). The competing demands of popularisation *vs* economy: Written language in the age of mass literacy. In T. Nevalainen and E. C. Traugott, eds., *The Oxford Handbook of the History of English*. Oxford: Oxford University Press, pp. 314–328.

Biber, D., Johansson, S., Leech, G., Conrad, S., and Finegan, E., eds. (1999). *Longman Grammar of Spoken and Written English*. Harlow: Pearson Education Limited.

Blevins, B., Rice, S., and Carpenter, R. (2015). Designing scholarly multimodal texts: A peer review process. *The Peer Review* (October 2015). Retrieved from http://thepeerreview-iwca.org/issues/issue-0/designing-scholarly-multimodal-texts-a-peer-review-process/ [Last accessed on 1 July 2021].

Blommaert, J. (2005). *Discourse: A Critical Introduction*. Cambridge: Cambridge University Press.

Bolter, J. D., and Grusin, R. (1999). *Remediation: Understanding New Media*. Cambridge: The MIT Press.

Bondi, M., Cacchiani, S., and Mazzi, D. (2015). Discourse in and through the media: Recontextualizing and reconceptualizing expert discourse. In M. Bondi, S. Cacchiani, and D. Mazzi, eds., *Discourse in and through the Media: Recontextualizing and Reconceptualizing Expert Discourse*. Newcastle upon Tyne: Cambridge Scholars Publishing, pp. 1–21.

Bonney, R., Cooper, C. B., Dickinson, J., Kelling, S., Phillips, T., Rosenberg, K. V., and Shirk, J. (2009). Citizen science: A developing tool for expanding science knowledge and scientific literacy. *BioScience*, 59(11), 977–984.

Boyd, D. M. (2002). *Faceted Identity: Managing Representation in a Digital World*. Unpublished master's thesis. Cambridge, MA: Massachusetts Institute of Technology.

Brandtzaeg, P. B., and Lüders, M. (2018). Time collapse in social media: Extending the context collapse. *Social Media + Society*, 4(1). https://doi.org/10.1177/2056305118763349

Bray, N. (2019). How do online news genres take up knowledge claims from a scientific research article on climate change? *Written Communication*, 36(1), 155–189.

Breeze, R. (2019). Continuity and change. Negotiating relationships in traditional and online peer review genres. In M. J. Luzón and C. Pérez-Llantada, eds., *Science Communication on the Internet: Old Genres Meet New Genres*. Amsterdam/ Philadelphia: John Benjamins, pp. 107–129.

Brezina, V., McEnery, T., and Wattam, S. (2015). Collocations in context: A new perspective on collocation networks. *International Journal of Corpus Linguistics*, 20(2), 139–173.

Brezina, V., Weill-Tessier, P., and McEnery, T. (2020). #LancsBox 5.x and 6.x [software]. Available at: http://corpora.lancs.ac.uk/lancsbox.

Briggs, C. L., and Bauman, R. (1992). Genre, intertextuality and social power. *Journal of Linguistic Anthropology*, 2(2), 131–172.

Büchi, M. (2016). Microblogging as an extension of science reporting. *Public Understanding of Science*, 26(8), 953–968.

Buehl, J. (2016). *Assembling Arguments: Multimodal Rhetoric and Scientific Discourse*. Columbia: University of South Carolina Press.

Burns, T. W., O'Connor, D. J., and Stocklmayer, S. M. (2003). Science communication: A contemporary definition. *Public Understanding of Science*, 12(2), 183–202. https://doi.org/10.1177/09636625030122004

Caliendo, G. (2012). The popularisation of science in web-based genres. In G. Caliendo and G. Bongo, eds., *The Language of Popularisation: Theoretical and Descriptive Models*. Bern: Peter Lang, pp. 101–132.

Carter-Thomas, S., and Rowley-Jolivet, E. (2017). Open science notebooks: New insights, new affordances. *Journal of Pragmatics*, 116, 64–76.

Casper, C. F. (2016). The online research article and the ecological basis of new digital genres. In A. G. Gross and J. Buehl, eds., *Science and the Internet: Communicating Knowledge in a Digital Age*. Amityville: Baywood, pp. 77–98.

Catenaccio, P. (2012). A genre-theory approach to the website: Some preliminary considerations. In S. Campagna, G. Garzone, C. Ilie, and E. Rowley-Jolivet, eds., *Evolving Genres in Web Mediated Communication*. Bern: Peter Lang, pp. 27–52.

Charles, M. (2015). Same task, different corpus: The role of personal corpora in EAP classes. In A. Leńko-Szymańka and A. Boulton, eds., *Multiple Affordances of Language Corpora for Data-Driven Learning*. Amsterdam: John Benjamins, pp. 131–154.

Cody, E. M., Reagan, A. J., Mitchell, L., Dodds, P. S., et al. (2015). Climate change sentiment on Twitter: An unsolicited public opinion poll. *PLoS ONE*, 10(8), e0136092. https://dx.plos.org/10.1371/journal.pone.0136092

Conrad, S., and Biber, D. (2001). *Variation in English: Multi-Dimensional Studies*. London: Longman.

Cox, L. (2015). *Are Graphical Abstracts Changing the Way We Publish?* Retrieved from   https://www.wiley.com/network/researchers/promoting-your-article/are -graphical-abstracts-changing-the-way-we-publish [Last accessed on 20 June 2020].

Coxhead, A. (2000). A new academic word list. *TESOL Quarterly*, 34, 213–238.

De Groot, E., Nickerson, C., Korzilius, H., and Gerritsen, M. (2016). Picture this: Developing a model for the analysis of visual metadiscourse. *Journal of Business and Technical Communication*. 30(2), 165–201.

Devitt, A. J. (2004). *Writing Genres*. Carbondale: Southern Illinois University.

Devitt, A. J. (2009). Re-fusing form in genre study. In J. Giltrow and D. Stein, eds., *Genres in the Internet: Issues in the Theory of Genre*. Philadelphia: John Benjamins Publishing Company, pp. 27–47.

Domingo, M., Jewitt, C., and Kress, G. (2015). Multimodal social semiotics: Writing in online contexts. In K. Pahl and J. Rowsel, eds., *The Routledge Handbook of Contemporary Literacy Studies*. London: Routledge, pp. 251–266.

Elleström, L. (2010). The modalities of media: A model for understanding intermedial relations. In L. Elleström, ed., *Media Borders, Multimodality and Intermediality*. London: Palgrave, pp. 11–48.

Engberg, J., and Maier, C. D. (2015). Exploring the hypermodal communication of academic knowledge beyond generic structure. In M. Bondi, S. Cacchiani, and D. Mazzi, eds., *Discourse in and through the Media: Recontextualizing and Reconceptualizing Expert Discourse*. Newcastle upon Tyne: Cambridge Scholars Publishing, pp. 46–62.

Facchinetti, R., and Palmer, F., eds. (2004). *English Modality in Perspective. Genre Analysis and Contrastive Studies*. Frankfurt/Berlin: Peter Lang.

Fahnestock, J. (1986). Accommodating science: The rhetorical life of scientific facts. *Written Communication*, 3(3), 275–296.

Fairclough, N. (1992). *Discourse and Social Change*. Cambridge: Polity Press.

Fairclough, N. (2003). *Analyzing Discourse. Textual Analysis for Social Research*. London/New York: Routledge.

Fecher, B., and Friesike, S. (2014). Open science: One term, five schools of thought. In S. Bartling and S. Friesike, eds., *Opening Science*. Cham: Springer, pp. 17–47.

Fletcher, W. H. (2002–2007). *KfNgram*. Annapolis: United States Naval Academy.

Flowerdew, J. (2002). Genre in the classroom: A linguistic approach. In A. Johns, ed., *Genres in the Classroom: Multiple Perspectives*. Mahwah: Lawrence Erlbaum, pp. 91–102.

Flowerdew, L. (2001). Corpus linguistics in ESP: A genre-based perspective. In A. Furness, G. H. Y. Wong, and L. Wu, eds., *Penetrating Discourse: Integrating Theory with Practice*. Hong Kong: Center for Language Education, Hong Kong University of Science and Technology, pp. 21–40.

Flowerdew, L. (2012). Corpus-based discourse analysis. In J. P. Gee and M. Handford, eds., *The Routledge Handbook of Discourse Analysis*. London/New York: Routledge, pp. 174–188.

Freadman, A. (1994). Anyone for tennis? In A. Freedman and P. Medway, eds., *Genre and the New Rhetoric*. London and New York: Taylor and Francis, pp. 43–66.

Freadman, A. (2002). Uptake. In R. M. Coe, L. Lingard, and T. Teslenko, eds., *The Rhetoric and Ideology of Genre: Strategies for Stability and Change*. Cresskill: Hampton Press, pp. 39–53.

Freedman, A., and P. Medway, eds. (1994). *Genre and the New Rhetoric*. London: Taylor and Francis.

Fryer, D. L. (2016). Cut and paste: Recontextualizing meaning-material in a digital environment. In S. Gardner and S. Alsop, eds., *Systemic Functional Linguistics in the Digital Age* (pp. 151–165). Sheffield: Equinox, pp. 151–165.

Genette, G. (1997). *Paratexts. Thresholds of Interpretation*. Foreword by R. Macksey. Translated by J. E. Lewin. Cambridge: Cambridge University Press.

Giltrow, J., and Stein, D. (2009). *Genres in the Internet: Issues in the Theory of Genre*. Amsterdam/Philadelphia: John Benjamins.

Gimenez, J., Baldwin, M., Breen, P., Green, J., Gutierrez, E., Paterson, R., Pearson, J., Percy, M., Specht, D., and Waddell, G. (2020). Reproduced, reinterpreted, lost: Trajectories of scientific knowledge across contexts. *Text & Talk*, 40(3), 293–324.

Gimenez, J., and Ernesto Roque Gutierrez. (2016). A pedagogy for multiple-audience writing in STEM disciplines: When research and pedagogy meet. Paper presented at the HEA: Horizons in STEM Higher Education, Leicester, UK: University of Leicester. 30th June–1st July

Goodman, S. N., Fanelli, D., and Ioannidi, J. P. A. (2016). What does research reproducibility mean? *Science Translational Medicine*, 8, 341ps12. https://doi .org/10.1126/scitranslmed.aaf5027

Gotti, M. (2014). Reformulation and recontextualisation in popularisation discourse. *Ibérica, Journal of the European Association of Languages for Specific Purposes*, 27, 15–34.

Gross, A. G. (1994). The roles of rhetoric in the public understanding of science. *Public Understanding of Science*, 3, 3–23.

Gross, A. G., and Buehl, J. (2016). *Science and the Internet: Communicating Knowledge in a Digital Age*. Amityville: Baywood.

Gross, A. G., and Harmon, J. E. (2016). *The Internet Revolution in the Sciences and Humanities*. Oxford: Oxford University Press.

Gruber, H., and Muntigl, P. (2005). Generic and rhetorical structures of texts: Two sides of the same coin? *Folia Linguistica*, XXXIX(1–2), 75–113.

Hafner, C. A. (2018). Genre innovation and multimodal expression in scholarly communication: Video methods articles in experimental biology. *Ibérica. Journal of the European Association of Languages for Specific Purposes*, 36, 15–42.

Halliday, M. A. K. (1978). *Language as Social Semiotic: The Social Interpretation of Language and Meaning*. London: Edward Arnold.

Harmon, J. E. (2019). At the frontiers of the online scientific article. In M. J. Luzón and C. Pérez-Llantada, eds., *Genres and Science in the Digital Age: Connecting Traditional and New Genres*. Amsterdam/Philadelphia: John Benjamins, pp. 19–40.

Herring, S. C. (2019). The co-evolution of computer-mediated communication and computer-mediated discourse analysis. In P. Bou-Franch and P. Garcés-Conejos Blitvich, eds., *Analysing Digital Discourse: New Insights and Future Directions*. London: Palgrave Macmillan, pp. 25–67.

Herring, S. C., Scheidt, L. A., Bonus, S., and Wright, E. (2004), Bridging the gap: A genre analysis of weblogs. Proceedings of the 37th Hawai'I International Conference on System Sciences (HICSS-37), Los Alamitos: IEEE Computer Society Press, pp. 1–11.

Hunston, S., and Francis, G. (1996). *Pattern Grammar: A Corpus-driven Approach to the Lexical Grammar of English*. Amsterdam/Philadelphia: John Benjamins.

Hunston, S., and Thompson, G., eds. (2000). Evaluation in texts. In *Authorial Stance and the Construction of Discourse*. Oxford: Oxford University Press.

Hyland, K. (2010). Constructing proximity: Relating to readers in popular and professional science. *Journal of English for Academic Purposes*, 9, 116–127.

Hyland, K. (2018). Narrative, identity and academic storytelling. *ILCEA* [Online], 31. Retrieved from https://journals.openedition.org/ilcea/4677#tocto1n7 [Last accessed on 15 November 2021].

Hyland, K., and Tse, P. (2004). Metadiscourse in academic writing. A reappraisal. *Applied Linguistics*, 25, 156–177.

Iedema, R. (2003). Multimodality, resemiotization: Extending the analysis of discourse as multi-semiotic practice. *Visual Communication*, 2(1), 29–57.

Jamieson, K. H., and Campbell, K. K. (1982). Rhetorical hybrids: Fusions of generic elements. *Quarterly Journal of Speech*, 69, 146–157.

Jamieson, K. M. (1975). Antecedent genre as rhetorical constraint. *Quarterly Journal of Speech*, 61, 406–415.

Jewitt, C. (2016). Multimodal analysis. In A. Georgakopoulou and T. Spilioti, eds., *The Routledge Handbook of Language and Digital Communication*. London: Routledge, pp. 69–84.

Jewitt, C., Bezemer, J., and O'Halloran, K. (2016). *Introducing Multimodality*. London: Routledge.

Johns, T. (1991). 'Should you be persuaded': Two samples of data-driven learning. *English Language Research Journal*, 4, 1–16.

Jones, R. H., and Hafner, C. A. (2012). *Understanding Digital Literacies: A Practical Introduction*.1st edition. Oxon: Routledge.

Jones, R. H., Chik, A., and Hafner, C. A. (2015). Introduction. Discourse analysis and digital practices. In R. H. Jones, A. Chik, and C. A. Hafner, eds., *Discourse and Digital Practices: Doing Discourse Analysis in the Digital Age*. London: Routledge, pp. 1–20.

Kelly, A. R. (2014). Hacking Science: Emerging Parascientific Genres and Public Participation in Scientific Research. Dissertation. Raleigh, NC: NCSU Institutional Repository. [Last accessed on 15 July 2021].

Kelly, A. R., and Maddalena, K. (2016). Networks, genres, and complex wholes: Citizen science and how we act together through typified text. *Canadian Journal of Communication*, 41(2), 287–303.

Kelly, A. R., and Miller, C. R. (2016). Intersections: Scientific and parascientific communication on the Internet. In A. Gross and J. Buehl, eds., *Science and the Internet: Communicating Knowledge in a Digital Age*. Amityville: Baywood, pp. 221–245.

Kim, H. J. (2000). Motivations for hyperlinking in scholarly electronic articles: A qualitative study. *Journal of the American Society for Information Science*, 51, 887–899.

Kress, G. (2010). *Multimodality: A Social Semiotic Approach to Contemporary Communication*. London: Taylor and Francis, Routledge.

Kress, G., and Van Leeuwen, T. (2001). *Multimodal Discourse: The Modes and Media of Contemporary Communication*. London: Arnold.

Kress, G., and Van Leeuwen, T. (2006). *Reading Images The Grammar of Visual Design*. London/New York: Routledge.

Kristeva, J. (1980), *Desire in Language: A Semiotic Approach to Literature and Art*. New York: Columbia University Press.

Kuehne, L., and Olden, J. (2015). Lay summaries needed to enhance science communication. *Proceedings of the National Academy of Sciences of the United States*, 112(12), 3585–3586. https://doi.org/10.1073/pnas.1500882112

Kwasnik, B. H., and Crowston, K. (2005). Genre of digital documents. *Introduction to the Special Issue of Information, Technology & People*, 18(2), 76–88.

Kwok, R. (2018). Lab notebooks go digital. *Nature*, 560, 269–270.

Lähdesmäki, S. (2009). Intertextual analysis of Finnish EFL textbooks: Genre embedding as recontextualization. In C. Bazerman, A. Bonini, and D. Figueiredo, eds., *Genre in a Changing World*. Fort Collins: The WAC Clearinghouse and Parlor Press, pp. 375–392.

Lang, A. (2014). Dynamic human-centered communication systems theory. *The Information Society*, 30(1), 60–70, DOI: 10.1080/01972243.2013.856364

Lazard, A., and Atkinson, L. (2015). Putting environmental infographics center stage: The role of visuals at the elaboration likelihood model's critical point of persuasion. *Science Communication*, 37(1), 6–33.

Lee, D. Y. W. (2008). Corpora and discourse analysis. New ways of doing old things. In V. K. Bhatia, J. Flowerdew, and R. H. Jones eds., *Advances in Discourse Studies*. London/New York: Routledge, pp. 86–99.

Leech, G., Hundt, M., Mair, C., and Smith, N. (2009). *Change in Contemporary English: A Grammatical Study*. Cambridge: Cambridge University Press.

Lemke, J. L. (1998). Multiplying meaning: Visual and verbal semiotics in scientific text. In J. Martin and R. Veel, eds., *Reading Science*. London: Routledge, 87–113.

Lemke, J. L. (2002). Travels in hypermodality. *Visual Communication*, 1(3), 299–325.

Lemke, J. L. (2006). Towards critical multimedia literacy: Technology, research, and politics. In M. McKenna, ed., *International Handbook of Literacy and Technology*. Vol II. Mahwah: Lawrence Erlbaum Associates, pp. 3–14.

Leppänen, S., and Kytölä, S. (2017). Investigating multilingualism and multisemioticity as communicative resources in social media. In M. Martin-Jones and D. Martin, eds., *Researching Multilingualism: Critical and Ethnographic Perspectives*. London: Routledge, pp. 155–171.

Lim, F. V., and Jiang, J. (2021). Towards a framework for language-visual-gestural intersemiosis: A multimodal analysis of TED talks. Paper presented at the 6th International Conference of Languages for Specific Purposes & Professional Communication. City University of Hong Kong, 3–5 June 2021.

Linell, P. (1998). Discourse across boundaries: On recontextualizations and the blending of voices in professional discourse. *Texts*, 18(2), 143–157.

Loi, C. K., and Evans, M. S. (2010). Cultural differences in the organization of research article introductions from the field of educational psychology: English and Chinese. *Journal of Pragmatics*, 42, 2814–2825.

Loroño-Leturiondo, M., and Davies, S. R. (2018). Responsibility and science communication: Scientists' experiences of and perspectives on public communication activities. *Journal of Responsible Innovation*, 5(2), 170–185.

Luzón, M. J. (2009). Scholarly hyperwriting: The function of links in academic weblogs. *Journal of the American Society for Information Science and Technology*, 60(1), 75–89.

Luzón, M. J. (2013). Public communication of science in blogs: Recontextualizing scientific discourse for a diversified audience. *Written Communication*, 30(4), 428–457.

Luzón, M. J. (2017). Connecting genres and languages in online scholarly communication: An analysis of research group blogs. *Written Communication*, 34(4), 441–471.

Luzón, M. J. (2020). Visual communication in online academic genres: An analysis of images on the websites of research groups. In M. Gotti, S. M. Maci, and M. Sala, eds., *Pathways: Knowledge Transfer and Knowledge Exchange in Academia*. Bern: Peter Lang, pp. 281–304.

Luzón, M. J., and Pérez-Llantada, C., eds. (2019). *Science Communication on the Internet: Old Genres Meet New Genres*. Amsterdam/Philadelphia: John Benjamins.

Luzón, M. J., and Pérez-Llantada, C. (2022). *Digital Genres in Academic Knowledge Production and Dissemination: Perspectives and Practices*. Bristol: Multilingual Matters.

Maier, C. D., and Engberg, J. (2018). Researchers' move from page to screen: Addressing the effects of the video article format upon academic user engagement and knowledge-building processes. In C. S. Guinda, ed., *Engagement in Professional Genres*. Amsterdam: John Benjamins, pp. 179–195.

Maier, C. D., and Engberg, J. (2019). The multimodal bridge between academics and practitioners in the Harvard Business Review's digital context: A multi-levelled qualitative analysis of knowledge construction. In M. J. Luzón and C. Pérez-Llantada, eds., *Science Communication on the Internet: Old Genres Meet New Genres*. Amsterdam/Philadelphia: John Benjamins, pp. 131–152.

Martinec, R., and Salway, A. (2005). A system for image-text relations in new (and old) media. *Visual Communication*, 4(3), 337–371.

Marwick, A., and boyd, d. (2011). I tweet honestly, I tweet passionately: Twitter users, context collapse, and the imagined audience. *New Media and Society*, 13, 96–113.

Mauranen, A. (2013). Hybridism, edutainment, and doubt: Science blogging finding its feet. *Nordic Journal of English Studies*, 13(1), 7–36.

Mauranen, A., Pérez-Llantada, C., and Swales, J. M. (2020). Academic Englishes: A standardised knowledge? In A. Kirkpatrick, ed., *The World Englishes Handbook*. 2nd edition. Abingdon/Oxon: Routledge, pp. 659–676.

Mavers, D. (2011). *The Remarkable in the Unremarkable: Children's Drawing and Writing*. New York: Routledge.

McEnery, T., and Hardie, A. (2011). *Corpus Linguistics*. Cambridge: Cambridge University Press.

Mehlenbacher, A. R. (2017). Crowdfunding science: Exigencies and strategies in an emerging genre of science communication. *Technical Communication Quarterly*, 26(2), 127–144.

Mehlenbacher, A. R. (2019a). Registered reports: An emerging scientific research article genre. *Written Communication*, 36(1), 38–67.

Mehlenbacher, A. R. (2019b). *Science Communication Online. Engaging Experts and Publics on the Internet*. Columbus: The Ohio State University Press.

Mehlenbacher, A. R., and Mehlenbacher, B. (2020). Rhetorical actors: Scientists and the social action of tweeting. In S. Auken and C. Sunesen, eds., *Genres of the Climate Debate*. Berlin: Mouton De Gruyter, pp. 177–191.

Miller, C. R. (1984). Genre as social action. *Quarterly Journal of Speech*, 70(2), 151–167.

Miller, C. R., and Fahnestock, J. (2013). Genres in scientific and technical rhetoric. *Poroi*, 9, 1: Article 12. https://doi.org/10.13008/2151-2957.1161

Miller, C. R., and Kelly, A. R., eds. (2017). *Emerging Genres in New Media Environments*. London: Palgrave Macmillan.

Millet, J. K., Tang, T., Nathan, L., Jaimes, J. A., Hsu, H. L., Daniel, S., and Whittaker, G. R. (2019). Production of pseudotyped particles to study highly pathogenic coronaviruses in a biosafety level 2 setting. *Journal of Visualized Experiments* (145), e59010. doi:10.3791/59010.

Molle, D., and Prior, P. (2008). Multimodal genre systems in EAP writing pedagogy: Reflecting on a needs analysis. *TESOL Quarterly*, 42(4). https://doi.org/10.1002/j.1545-7249.2008.tb00148.x

Motta-Roth, D. (2009). Popularização da ciência como prática social e discursiva. In D. Motta-Roth and M. E. Giering, eds., *Discursos de popularização da ciência. Hipers@beres*. Santa Maria: PPGL Editores, pp. 131–195.

Motta-Roth, D., and Scherer, A. S. (2016). Science popularization: Interdiscursivity among science, pedagogy, and journalism. *Bakhtiniana*, 11(2), 171–194.

Munafò, M., Nosek, B., Bishop, D., et al. (2017). A manifesto for reproducible science. *Nature Human Behaviour*, 1, 0021. https://doi.org/10.1038/s41562-016-0021

Myers, G. (2003). Discourse studies of scientific popularization: Questioning the boundaries. *Discourse Studies*, 5(2), 265–279.

Nissenbaum, H. (2009). *Privacy in Context: Technology, Policy, and the Integrity of Social Life*. Stanford: Stanford University Press.

O'Halloran, K. L. (2005). *Mathematical Discourse: Language, Symbolism and Visual Images*. London/New York: Continuum.

O'Halloran, K. L. (2009). Multimodal analysis and digital technology. In A. Baldry and E. Montagna, eds., *Interdisciplinary Perspectives on Multimodality: Theory and Practice*. Campobasso: Palladino, pp. 21–34.

O'Halloran, K. L., and Lim, F. V. (2014). Systemic functional multimodal discourse analysis. In S. Norris and C. Maier, eds., *Texts, Images and Interactions: A Reader in Multimodality*. Berlin: Mouton de Gruyter, pp. 137–154.

O'Halloran, K. L., Tan, S., and E, M. K. L. (2015). Multimodal analysis for critical thinking. *Learning, Media and Technology*, 42(2) 147–170.

O'Halloran, K. L., Tan, S., and Wignell, P. (2019). SFL and multimodal discourse analysis. In G. Thompson, W. L. Bowcher, L. Fontaine, and J. Y. Liang, eds., *The Cambridge Handbook of Systemic Functional Linguistics*. Cambridge, UK: Cambridge University Press, pp. 433–461.

Orlikowski, W. J., and Yates, J. (1994). Genre repertoire: The structuring of communicative practices in organizations. *Administrative Science Quarterly*, 39(4), 541–574.

Orpin, D. (2019). #Vaccineswork: Recontextualizing the content of epidemiology reports on Twitter. In M. J. Luzón and C. Pérez-Llantada, eds., *Science Communication on the Internet. Old Genres Meet New Genres*. Amsterdam/Philadelphia: John Benjamins, pp. 173–194.

Østergaard, S., and Bundgaard, P. F. (2015). The emergence and nature of genres: A social-dynamic account. *Cognitive Semiotics*, 8(2), 97–127.

Owen, R., Macnaghten, P., and Stilgoe, J. (2012). Responsible research and innovation: From science in society to science for society, with society. *Science and Public Policy*, 39(6), 751–760.

Page, R. (2012). The linguistics of self-branding and micro-celebrity in Twitter: The role of hashtags. *Discourse & Communication*, 6(2), 181–201.

Paltridge, B. (2012). *Discourse Analysis: An Introduction*. London: Bloomsbury Academic.

Pasquali, M. (2007). Video in science. Protocol videos: The implications for research and society. *EMBO Reports*, 8(8), 712–716.

Paulus, T. M., and Roberts, K. R. (2018). Crowdfunding a real-life superhero: The construction of worthy bodies in medical campaign narratives. *Discourse, Context & Media*, 21, 64–72.

Pauwels, L. (2006). *Visual Cultures of Science: Rethinking Representational Practices in Knowledge Building and Science Communication*. Lebanon: Dartmouth College Press.

Pérez-Llantada, C. (2013). The Article of the Future: Strategies for genre stability and change. *English for Specific Purposes*, 32(4), 221–235.

Pérez-Llantada, C. (2015). Genres in the forefront, languages in the background: The scope of genre analysis in language-related scenarios. *Journal of English for Academic Purposes*, 19, 10–21.

Pérez-Llantada, C. (2019). Ecologies of genres and an ecology of languages of science: Current and future debates. In D. R. Gruber and L. C. Walsh, eds., *The Routledge Handbook of Language and Science*. New York: Routledge, pp. 361–374.

Pérez-Llantada, C. (2021a). Grammar features and discourse style in digital genres: The case of science-focused crowdfunding projects. *Revista Signos. Estudios de Lingüística*, 54(105), 73–96.

Pérez-Llantada, C. (2021b). *Research Genres across Languages. Multilingual Communication Online*. Cambridge: Cambridge University Press.

Plastina, A. F. (2017). Professional discourse in video abstracts: Re-articulating the meaning of written research article abstracts. In G. Garzone, P. Catenaccio, K. Grego, and R. Doerr, eds., *Specialised and Professional Discourse across Media and Genres*. Milano: Ledizioni, pp. 57–74.

Prem, E., Sanz, F. S., Lindorfer, M., Lampert, D., and Irran, J. (2016). *Open Digital Science*. Final study report. Retrieved from https://ec.europa.eu/digital-single -market/en/news/open-digital-science-final-study-report [Last accessed on 20 January 2021].

Prior, P. A. (2009). From speech genres to mediated multimodal genre systems: Bakhtin, Voloshinov, and the question of writing. In C. Bazerman, A. Bonini, and D. Figueiredo, eds., *Genre in a Changing World*. WAC Clearinghouse and Parlor Press, pp. 17–34.

Prior, P. A. (2013). Multimodality and ESP research. In B. Paltridge and S. Starfield, eds., *The Handbook of English for Specific Purposes*. 1st edition. Boston: Wiley-Blackwell, pp. 519–534.

Prior, P., and Hengst, J. (2010). Introduction: Exploring semiotic remediation. In P. Prior and J. Hengst, eds., *Exploring Semiotic Remediation as Discourse Practice*. New York: Palgrave, pp. 1–23.

Puschmann, C. (2014). (Micro)blogging science? Notes on potentials and constraints of new forms of scholarly communication. In S. Friesike and S. Bartling, eds., *Opening Science*. New York: Springer, pp. 89–106.

Reid, G. (2019). Compressing, expanding, and attending to scientific meaning: Writing the semiotic hybrid of science for professional and citizen scientists. *Written Communication*, 36(1), 68–98.

Reid, G., and Anson, C. M. (2019). Public- and expert-facing communication: A case study of polycontextuality and context collapse in Internet-mediated citizen science. In M. J. Luzón and C. Pérez-Llantada, eds., *Science Communication on the Internet. Old Genres Meet New Genres*. Amsterdam/Philadelphia: John Benjamins, pp. 219–238.

Römer, U. (2008). Identification impossible?: A corpus approach to realisations of evaluative meaning in academic writing. *Functions of Language*, 15(1), 115–130.

Römer, U. (2010). Establishing the phraseological profile of a text type. The construction of meaning in academic book reviews. *English Text Construction*, 3(1), 95–119.

Roque, G. (2017). Rhetoric, argumentation, and persuasion in a multimodal perspective. In A. Tseronis and C. Forceville, eds., *Multimodal Argumentation and Rhetoric in Media Genres*. Amsterdam/Philadelphia: John Benjamins, pp. 25–50.

Ross-Hellauer, T. (2017). What is open peer review? A systematic review [version 2; peer review: 4 approved]. *F1000Research*, 6, 588.

Rowley-Jolivet, E., and Carter-Thomas, S. (2019). Scholarly soundbites: Audiovisual innovations in digital science and their implications for genre evolution. In M. J. Luzón and C. Pérez-Llantada, eds., *Science Communication on the Internet: Old Genres Meet New Genres*. Amsterdam: John Benjamins Publishing Company, pp. 81–106.

Salmose, N., and Elleström, L., eds. (2020). *Communication across Media Borders*. London/New York: Routledge.

Sancho-Guinda, C. (2019). Promoemotional science? Emotion and intersemiosis in graphical abstracts. In J. L. Mackenzie and L. Alba Juez, eds., *Emotion in Discourse*: Amsterdam/Philadelphia: John Benjamins, pp. 357–386.

Sarcina, A. (2019) Open Science: A review of definitions with a regional perspective. *Impakter*, December 6. https://impakter.com/open-science-a-review-of -definitions-with-a-regional-perspective/

Schryer, C. F. (1994). The lab vs. the clinic: Sites of competing genres. In A. Freedman and P. Medway, eds., *Genre and the New Rhetoric*. London: Taylor and Francis, pp. 105–124.

Schulson, M. (2018). Science's 'reproducibility crisis' is being used as political ammunition. *Wired*, 20 April 2018. Retrieved from https://www.wired.com/ story/sciences-reproducibility-crisis-is-being-used-as-political-ammunition/ [Last accessed on 10 April 2020].

Science Europe (2017). The rationales of open science: Digitalisation and democratisation in research. *Science Europe High-Level Workshop*, September 14, 2017, Berlin. Retrieved from https://www.dfg.de/download/pdf/dfg_magazin /veranstaltungen/internationales/rationales_of_open_science.pdf [Last accessed on 14 November 2021].

Scollon, R. (2008). Discourse itineraries: Nine processes of resemiotization. In V. K. Bhatia, J. Flowerdew, and R. H. Jones, eds., *Advances in Discourse Studies*. London/New York: Routledge, pp. 233–244.

Scott, M. (2008). *Wordsmith Tools 5*. Liverpool: Lexical Analysis Software.

Scotto di Carlo, G. (2014). The role of proximity in online popularisations: The case of TED talks. *Discourse Studies*, 16(5), 591–606. https://doi.org/10.1177 /1461445614538565

Shaklee, P. M. (2014). Data in brief: Making your data count. *Data in Brief*, 1, 5–6. https://doi.org/10.1016/j.dib.2014.09.001

Shema, H., Bar-Ilan, J., and Thelwall, M. (2012). Research blogs and the discussion of scholarly information. *PloS ONE*, 7(5), e35869.

Sinclair, J. McH. (2008). The phrase, the whole phrase and nothing but the phrase. In S. Granger and F. Meunier, eds., *Phraseology: An Interdisciplinary Perspective*. Amsterdam: John Benjamins, pp. 407–410.

Sindoni, M. G. (2013). *Spoken and Written Discourse in Online Interactions: A Multimodal Approach*. London/New York: Routledge.

Smart, G., and Falconer, M. (2019). The representation of science and technology in genres of Vatican discourse: Pope Francis's encyclical Laudato Si' as a case study. In M. J. Luzón and C. Pérez-Llantada, eds., *Science Communication on the Internet. Old Genres Meet New Genres*. Amsterdam/Philadelphia: John Benjamins, pp. 195–218.

Spicer, S. (2014). Exploring video abstracts in science journals: An overview and case study. *Journal of Librarianship and Scholarly Communication*, 2(2), eP1110.

Spinuzzi, C. (2003). *Tracing Genres through Organizations: A Sociocultural Approach to Information Design*. Cambridge, MA: MIT Press.

Spinuzzi, C. (2004). *Describing Assemblages: Genre Sets, Systems, Repertoires, and Ecologies. Computer Writing and Research Lab*. White Paper Series: #040505-2. Austin: Digital Writing and Research Lab.

Spinuzzi, C., and Zachry, M. (2000). Genre ecologies: An open-system approach to understanding and constructing documentation. *ACM Journal of Computer Documentation*, 24(3), 169–181. https://doi.org/10.1145/344599.344646

Stubbs, M. (2001). *Words and Phrases: Corpus Studies in Lexical Semantics*. Oxford: Blackwell Publishers.

Swales, J. M. (1990). *Genre Analysis. English in Academic and Research Settings*. 1st edition. Cambridge: Cambridge University Press.

Swales, J. M. (1993). Genre and engagement. *Revue Belge de Philologie et de l'histoire*, 71, 687–698.

Swales, J. M. (1996). Occluded genres in the academy: The case of the submission letter. In E. Ventola and A. Mauranen, eds., *Academic Writing: Intercultural and Textual Issues*. Amsterdam: John Benjamins, pp. 45–58.

Swales, J. M. (2004). *Research Genres: Explorations and Applications*. Cambridge: Cambridge University Press.

Swales, J. M., and Feak, C. B. (1995). From information transfer to data commentary. *The Journal of TESOL*, 2(2), 79–92.

Swales, J. M., and Feak, C. B. (2009). *Abstracts and the Writing of Abstracts*. Ann Arbor: The University of Michigan Press.

Tachino, T. (2012). Theorizing uptake and knowledge mobilization: A case for intermediary genre. *Written Communication*, 29(4), 455–476.

Tandon, A. (2019). Climate change could cause 'irreversible impacts' to lake ecosystems. Retrieved from https://www.carbonbrief.org/ [Last accessed on 24 October 2021]

Thelwall, M. (2003). What is this link doing here? Beginning a fine grained process of identifying reasons for academic hyperlink creation. *Information Research*, 8(3). Retrieved from http:// informationr.net/ir/8-3/paper151.html [Last accessed on 15 December 2020]

Thelwall, M., and Kousha, K. (2015). ResearchGate: Disseminating, communicating, and measuring scholarship? *Journal of the Association for Information Science and Technology*, 66(5), 876–889.

Trench, B. (2008). Internet: Turning science communication inside-out? In M. Bucchi and B. Trench, eds., *Handbook of Public Communication of Science and Technology*. New York: Routledge, pp. 185–198.

Tribble, C. (2001). Small corpora and teaching writing. Towards a corpus-informed pedagogy of writing. In M. Ghadessy, A. Henry and R. L. Roseberry, eds., *Small Corpus Studies and ELT: Theory and Practice*. Amsterdam, Philadelphia: John Benjamins, pp. 381–408.

Tseronis, A., and Forceville, C. (2017). Argumentation and rhetoric in visual and multimodal communication. In A. Tseronis and C. Forceville, eds., *Multimodal Argumentation and Rhetoric in Media Genres*. Amsterdam: John Benjamins, pp. 1–24.

Van Dijk, T. A. (2014). *Discourse and Knowledge: A Sociocognitive Approach*. Cambridge: Cambridge University Press.

Van Leeuwen, T. (2005). Multimodality, genre and design. In S. Norris and R. H. Jones, eds., *Discourse in Action: Introducing Mediated Discourse Analysis*. London: Routledge, pp. 73–94.

Van Leeuwen, T. (2008). *Discourse and Practice: New Tools for Critical Discourse Analysis*. Oxford: Oxford University Press.

Vardell, E. (2015). JoVE: The journal of visualized experiments. *Medical Reference Services Quarterly*, 34(1), 88–97. https://doi.org/10.1080/02763869.2015.986795

White, P. R. R. (2003). Beyond modality and hedging: A dialogic view of the language of intersubjective stance. *Text and Talk*, 23(2), 259–284. https://doi .org/10.1515/text.2003.011

Woolway, et al. (2021). Phenological shifts in lake stratification under climate change. *Nature Communications* 12, 2318. https://doi.org/10.1038/s41467-021 -22657-4

Wynne, B. (2006). Public engagement as a means of restoring public trust in science: Hitting the notes, but missing the music? *Community Genetics*, 9(3), 211–220.

Yang, W. (2017). Audioslide presentations as an appendant genre: Key words, personal pronouns, stance and engagement. *ESP Today*, 5(1), 24–45. https://doi .org/10.18485/esptoday.2017.5.1.2

Yates, J., and Orlikowski, W. J. (1992). Genres of organizational communication: A structurational approach to studying communication and media. *Academy of Management Review*, 17(2), 485–510.

Yates, S. J., and Sumner, T. R. (1997). Digital genres and the new burden of fixity. *Proceedings of the Thirtieth Hawaii International Conference on System Sciences*, 6(6), 3–12.

Zappavigna, M. (2011). Ambient affiliation: A linguistic perspective on Twitter. *New Media and Society*, 13(5), 788–806.

Zappavigna, M. (2017). Twitter. In Ch. Hoffmann and W. Bublitz, eds., *Pragmatics of Social Media*. Berlin/Boston: De Gruyter Mouton, pp. 201–224.

Index

For Product Safety Concerns and Information please contact our EU
representative GPSR@taylorandfrancis.com
Taylor & Francis Verlag GmbH, Kaufingerstraße 24, 80331 München, Germany

www.ingramcontent.com/pod-product-compliance
Lightning Source LLC
Chambersburg PA
CBHW061438180526
45170CB00004B/1467